# Industrial Maintenance

T0179053

# Industrial Maintenance
## Techniques, Stories, and Cases

José Baptista

**CRC Press**
Taylor & Francis Group
Boca Raton  London  New York

CRC Press is an imprint of the
Taylor & Francis Group, an **informa** business

This book was previously published in 2016 in Portuguese by Lura Editoração Eletrônica Ltda, as Manutenção Industrial: Técnicas, contos e causos, by José Antonio Baptista.

CRC Press
Taylor & Francis Group
6000 Broken Sound Parkway NW, Suite 300
Boca Raton, FL 33487-2742

First issued in paperback 2020

© 2020 by Taylor & Francis Group, LLC
CRC Press is an imprint of Taylor & Francis Group, an Informa business

No claim to original U.S. Government works

ISBN-13: 978-0-367-34115-2 (hbk)
ISBN-13: 978-0-367-77659-6 (pbk)

**Visit the Taylor & Francis Web site at**
**http://www.taylorandfrancis.com**

**and the CRC Press Web site at**
**http://www.crcpress.com**

*For Rosana, my eternal beloved wife and companion,
and for Alexandre and Larissa, my dear children, those
who due to my work in maintenance have been deprived
of my company on many Sundays and holidays.*

*Many a time, their sleep has been interrupted by the frequent
ringing of the phone at inconvenient hours, particularly during
the night, to discuss some problem with me in several companies
that I worked for or calling me to repair some equipment, which
the tradesmen on duty was not able to fix within a reasonable
time given by the person responsible for production.*

# Contents

# Foreword

In 2003, I was responsible for the Automation division of a multinational company operating in oil, gas, chemical, and petrochemical industry segments. After about 3 years of hard work to recover the performance of this operation, the company started reporting very good financial and operational results. In one of the typical changes that occurs in large companies, I was informed that the service operations of a company, which was acquired few years ago, would report their results, related to the above industry segments, and therefore it was necessary for me to understand the reported numbers.

The service company focused on industrial maintenance outsourcing, with approximately 3,000 employees, internally called "Full Service." In addition, it provided technical assistance services in several areas spread across its various business units, which were known collectively as Field Service.

The Full-Service operation was going through a troubled phase and, quickly, what would have been a simple evaluation absorbed all my time and dedication to help the company solve various problems.

So, during a meeting with the President of the company, what was communicated to me left me wondering if I should celebrate or be angry: "We had a discussion with our people at Headquarters in Zurich, and we concluded that we should centralize all services operations—Full Service and Field Service—from all areas in to a single operation, and we want you as the new area manager."

Although I was happy for the company's consideration in believing that I would be able to help solve that problem, it came to my mind: "But why Service?" I am an electrical engineer, with a specialization in electronics, graduated in artificial intelligence, with years of working experience in automation. Working with something with a lower level of technology than I was used to let me down.

Yes, that is the typical ignorance and prejudice that many managers with good training in all types of companies have when we talk about services, and I did not escape the rule.

I began to get involved in this exercise and quickly, what at the beginning seemed the end of the line of engineering began to emerge as an area of high complexity involving a great deal of science and technology.

Several acronyms and concepts began to emerge before me: MTBF, MTTR, RCA, FMECA, CMMS, RCM, FMEA, OEE, TPM and LCC, among others.

It was at that moment that I had the opportunity to meet some of the major maintenance specialists in Brazil and around the world. For all of them, I have great admiration and respect. Among them, one had a special characteristic that caught my attention at the beginning, José Baptista. Yes, with this "p" in the middle of his last name, which I insisted writing incorrectly. Baptista was a rare blend of a great manager who was also a scientist and a teacher. He became the consultant of all the problems, of all the areas in all markets.

He certainly helped me understand why maintenance is poorly known, poorly managed, and discriminated against by most companies and ignorant managers, including myself.

As industrial maintenance outsourcing is one of our businesses, we were constantly assessing the maintenance processes of companies of the most diverse sectors. In this way, I was able to see in practice the lack of competence and appropriate focus in this area by almost all companies, national or multinational, in any market segment.

It was not uncommon to meet managers, with a few ready-to-effect phrases, trying to demonstrate dominance and knowledge that they did not possess. It also wasn't uncommon to meet directors starting the conversation and then transferring all discussions to their strong maintenance man who was trustworthy, hardworking, always recognized for acting as a fire extinguisher, and learned from the experience of day-to-day management, but without the knowledge necessary to lead the management of the assets to a level of world excellence. The losses in general in this area are high and unknown.

There is generally a strange appreciation for the "fire extinguisher" professional and, maybe, this is the reason why asset management does not evolve within these companies. It is essential that these managers acquire competence and the ability to communicate, without interlocutors, with executives who make the company's strategic decisions.

Baptista experienced, as an executive and a consultant specializing in asset management, the most diverse types of problems in this area in Brazil and in more than 30 countries around the world.

Surely, reading this book will enable you, from any area of activity, to be able to speak with fluency on this subject, which is so common and at the same time understood very little by most companies. As a Brazilian, I am proud to have, in our country, one professional with this level of competence and I am even more proud of having him as my teacher.

Good reading!

**Wilson Monteiro Junior**
*ABB Country Managing Director Peru*

# Preface

It's been said that over the past 40 years, industrial maintenance has changed more than any other management discipline. From my experience, I believe that to be true.

Advances in technology have increased equipment reliability, subsequently reducing maintenance costs and increasing the time between overhauls. This consequent reduction in maintenance has seen new techniques emerge. For example, Reliability-Centered Maintenance (RCM), Root Cause Analysis (RCA), Failure Mode and Effect Analysis (FMEA) and Life Cycle Cost (LCC).

Additionally, advances in information and communications technology such as Industrial Internet have enabled several innovations in maintenance such as predictive techniques and remote live monitoring of connected equipment.

Despite the aforementioned developments and innovations in maintenance, I've witnessed many companies of different industrial sectors around the world still work predominantly in the reactive run-to-fail maintenance mode.

The main motivation for writing this book has been my determination to demonstrate that operating in a purely reactive mode is inefficient and it's possible to transition into operating predominantly in the more efficient proactive mode.

I didn't want to write just another purely technical book. I understand that there are great technical books available on this subject, and I certainly have nothing to add to what already exists. However, I'm sure my book can add practical guidelines to the existing literature on the basis of my many years of experience in this area and from traveling around the world evaluating and auditing maintenance management systems, training people and supporting maintenance departments.

I truly believe that the experiences I will share with my readers will help them understand the essence of maintenance and that it is one of the most important activities for companies to achieve their strategic objectives.

So, I invite readers to follow me on this tour through the world of maintenance. I promise to share all the important things I have learned in all those years which if I had known years ago would have saved many hassles, equipment downtime, out-of-office hour calls, hard work on Sundays and holidays, and all the stress associated with those nasty situations.

This book will also review other important related subjects such as concepts and definitions of maintenance and reliability; how to develop maintenance plans using RCM and FMEA concepts; the importance of basic maintenance, maintenance planning and scheduling; how Sherlock Holmes can help you to understand and efficiently perform RCA, spare parts management, performance indicators, and address one of the most important topics that is frequently underestimated, which is the maintenance professionals and how human error influences maintenance.

# Author

**José Baptista** has 40 years of experience in the field of maintenance and reliability engineering, including 15 years at ABB, where he held several positions, highlighting ABB Full Service® site manager, consulting and engineering manager in Brazil. He was also the regional manager in North America and global reliability technology manager.

He has traveled to over 30 countries, doing technical trainings, assessing maintenance and reliability organizations, and supporting ABB Full Service® sites.

Before joining ABB, José worked as a maintenance manager at Bandag (now Bridgestone Bandag, LLC), as a process and project engineering manager at Fairchild Semiconductor and as the head of electrical and instruments maintenance in Pirelli, when he started his career.

He received B.S. degree from Faculdade de Engenharia Industrial in Brazil with a B.S. in Electrical Engineering in 1979.

He is a Certified Maintenance and Reliability Professional (CMRP) by SMRP in USA and a Certified Asset Management Assessor (CAMA) by World Partners in Asset Management (WPiAM).

In 2016, he started his own consulting company, along with his wife Rosana, named J&R Consultoria em Gestão de Manutenção e Confiabilidade Ltda. providing consultancy services and training to customers worldwide, such as ABB, Cenibra, Eaton, Leadec Industrial Services, Manserv, MODEC, Neste, Suzano, Voltalia and VOEE Services, among others.

# 1 The Sad Story of Antonio
## *Another Victim of the Reactive Maintenance Curse*

From my experience, in most cases, to work in the maintenance department is not an option but a lack of choice. When I started working in this area, I heard many people saying that maintenance is "the corner where the child cries and the mother does not listen!"—a phrase which itself illustrates the current thinking.

My case was no different, I majored in electrical engineering with a specialization in electronics, and my dream was to work in research or projects.

I had done an internship with a researcher in the college where I studied and was very excited about the job, but there was no way I was going to be made effective. Needing to work to share the expenses of my family, which had recently started, with my wife, I did not have much time to choose what to do, so I willingly accepted the challenge of being the head of the maintenance department in a tire plant. Two months after I graduated as an engineer, I started my career in industrial maintenance, and it was there where I spent most of my career.

To start talking about maintenance, there is nothing better than telling a story that illustrates a situation, unfortunately, very common in many industrial plants around the world:

The production supervisor, too fat and perspiring profusely, with red cheeks and panting, revealing the heat and his anger, opened the maintenance supervision office door with a bump already raging with all present:

"You are 'sawing the foot of my stool!' That damned line stopped again!"

Before anyone could observe any reaction, the uncontrolled supervisor continued to shout:

"This is the third time that the production line stopped in less than 24 hours. This is unacceptable! I will not be able to meet the production schedule, so, be prepared, because this is going south to everyone! Write it down: if the production line continues without reaching the goals, my neck will be on the line, I can even lose this damn job, but I'll take a lot of people with me. Mark my words!"

As soon as the supervisor left the room, slamming the door shut, the phone rang; it was the maintenance manager calling his subordinates for an immediate meeting, in order to discuss the production line stops since he needed to urgently deliver a report to his boss, the plant manager, justifying all the production losses caused by equipment breakdown.

In fact, the situation described was not new; on the contrary, it was a recurring nightmare that had become routine in that place; the production supervisor always exasperated, on the verge of a nervous breakdown and the maintenance personnel, cornered, doing everything they could and knew to remedy the situation, which, totally out of control, only got worse every day.

Overtime and maintenance costs went beyond the budget set for the department to a point where the industrial manager, at a meeting of budget monitoring, said to the maintenance manager, "this department of yours cost me much more money than a French lover does."

And so the meeting ended, and the maintenance manager became the target of jokes from other members of management, always being compared to a 'French lover'.

The members of the maintenance crew were all dejected and downcast by being judged as incompetent and villains of the company. One day, in this atmosphere of despair, the maintenance manager was approached secretly by one of the maintenance mechanics:

"Mr. Antonio, I really need to talk to you," said the mechanic in a low and discrete tone.

The maintenance manager just thought it was another of his employees that was resigning, since this was another fact that was becoming routine. Many maintenance workers couldn't handle the pressure and the terrible work environment, and were simply changing jobs. It was even said that people have agreed to a lower salary in another company just to be rid of that hell.

"Go ahead, Mariano, speak freely. By the look in your face I believe I know what the subject of our conversation will be!"

"We can't talk here Mr. Antonio," replied the mechanic timidly, without facing the manager – "This a private conversation."

"All right, you can go anytime to my office."

"I don't want talk there too, it would be better to talk somewhere else out of here."

"My God, the mystery!", said the manager, visibly losing patience. "Can't you brief me on the subject? You know how I've been busy, especially with these damned machines that won't stop breaking!"

"Can't tell you anything sir…," answered the mechanic in a reticent way, facing the filthy factory floor.

"All right then, let's meet at John's bar. Can you go there when we leave here today?"

"I prefer to meet you in the Cathedral's square, where I usually get off the company's bus. I will be sitting on a bench waiting for you. Could that work?"

The manager scratched his head and wearily agreed: "All right. I'll meet you at the Cathedral's square."

When Antonio arrived in the square, he noticed that Mariano was already sitting on one of the concrete benches, looking around impatiently.

He approached the mechanic and started a rushing conversation:

"Come, Mariano, spit it out at once what you have to tell me, because I want to go home soon."

"Mr. Antonio, I know that the situation of the maintenance department is very complicated and that we are not keeping this factory running. I've been thinking about it; we have done everything and still we can't solve the problem…."

Exasperated, the manager interrupted:

"For God's sake Mariano! You do all this mystery to come and talk about it?"

Unfazed by the boss' reprimand, the mechanic continued:

"You know Mr. Antonio, I've seen a lot in my life and if I tell, nobody will believe it – and with serious expression added – there are things that reason does not explain; supernatural things. Don't you think everything is going on at the maintenance department is very strange?"

"All right, I agree that we are living in a phase of problems but where are you going with this?", replied the manager.

The mechanic looked around and said quietly:

"I think there is something 'done' to us."
   "I don't understand. What do you mean?"
   "I mean they did a 'job'. Understood?"

The manager said laughing:

"You mean like voodoo, Mariano?"
   "Something like that and, you know Mr. Antonio, this is very serious; I have seen many people fall into disgrace because of a 'job' well done."
   "Who would do this to us? Why?", asked the still incredulous manager.
   "You fired so many people because of the many maintenance headcount reductions already. You can be sure that several former teammates left hurt and many people would be willing to take revenge. You never know who you're messing with...."
   "Well Mariano cut the crap. If this was everything you had to tell me, good night!"

With that said, the manager turned and walked quickly away from the square, leaving the mechanic alone, still sitting on the bench with a disconsolate look on his face.

The next day began with an emergency meeting convened by the plant manager: they would need to work on Sunday to recover the production losses and meet the schedule; otherwise, the product quantity would be short to meet customer requests.

And the tense meeting ended up with a message to the maintenance manager in a threatening tone:

"Listen to me Antonio, tell your staff that we'll work on Sunday, paying overtime to compensate for the maintenance department incompetence and therefore, the production lines cannot stop. This is the last chance we have to catch up. Is that clear?"
   "Yes sir."

Sunday morning, the phone at the maintenance manager's house rang early and he, still stunned, was listening to a squeaky and well-known voice on the other side:

"Mr. Antonio, this is Jurandir, maintenance supervisor of A shift, I have bad news; the finishing line stopped because the bearing of the main shaft burst, and worse, we do not have spare parts in stock. The production sent back home the operators because we have no estimate of when we can repair the line."

Momentarily the phone went silent.

"Mr. Antonio, can you hear me?"

"I do... I hear you perfectly. Are you sure that we do not have this bearing in stock?"

"The system indicates that we have 2 pieces, but on the shelf there's nothing... you know how it is, the shift personnel take the material and do not register in the system because the storekeeper was fired by yourself to reduce costs. It's a lot of pressure, there's no time for anything..."

"But why are we changing so many bearings?"

"This I do not know, sir, since the staff doesn't fulfill the work orders right and also, no one else does the maintenance report."

"Jurandir, please don't let the operators go home. I'll try to make contact with the supplier of the bearings to see if I can get some on an emergency basis so we can put the equipment into operation."

"The staff is gone, the bus just left."

Monday morning, in a meeting at the plant management room, the manager's shouts could be heard in the hallway:

"Antonio, I told you so. We can no longer trust you and your team; when we need it the most, the machines break down. I want you to tell me what do we need to do to have reliable equipment? You have until Friday at noon to give me the answer. Remember, my patience is over!"

Antonio left disconsolate, seriously considering resigning but also thinking about the house they recently bought. And the children's school? How long it would take to get another job? How could he provide for his family without his salary?

Another of his thoughts was towards the frustration and vexation in resigning his post after failing to keep the equipment running and meeting the production schedule; in fact, that would result in a incompetence certificate.

"I won't give up!", he thought and took a deep breath; the way now was to find the exit – find a solution to the maintenance of that plant.

What to do: hire consultants? Take courses on maintenance?

He is running against the clock, and certainly wouldn't have time to prepare a foolproof plan overnight; in fact, he didn't know where to start.

While walking through the factory, lost in his thoughts, he found Mariano, the same one he left at the square's bench a few days ago. Suddenly he had an idea, perhaps more by despair than by logic and directly addressed the mechanic:

"Mariano, I'd like talk again about that subject that we were discussing. Could you come to my office now?"

"But now...," stammered the mechanic.

"Immediately."

As he closed the room door, Antonio asked Mariano:

"You told me about a 'job' against us and I ask you what we can do to get rid of this curse that befell the machines?"

"Yes, Mr. Antonio that was just about what I wanted to talk that day," and the mechanic continued talking in very low tone, as if telling a secret: "You need to talk to a 'preto velho' (old black man)[1] I know, who has helped many people."

"Preto velho?", asked the manager.

"Yes. He is a holy man who can help, but you have to have faith, do what he says, and go where he goes."

"Does he charge money? Do you really think I can send a purchase order to cover the expenses with a macumba-man?"

"You won't need. He doesn't charge anything. If the problem is solved you can donate whatever you want for a nursery in the neighborhood where he lives."

"I'll think about it…," replied the manager reticently.

The mechanic was heading to the office door when the phone rang. The manager answered and remained silent, pale, just listening to what the person on the other end was saying. Once off the phone, he walked quickly to the room's door, preventing the mechanic leaving, saying:

"I thought better, please tell me how I can talk to this man you mentioned. What is his name?"

"He is known as Father Jau. I'll arrange so that you can meet him."

As soon as he said that, the mechanic quickly left the room, leaving the manager immersed in thoughts; the absurdity of the situation: many years of study and efforts to have to look for a healer to help and solve a technical problem; he must be dreaming or rather having a nightmare.

The same way it was a long shot to meet with Father Jau, he also knew he had no plan in mind and much less time to find any technical solution that could stabilize the plant situation, reducing the anger of the other managers, who felt their jobs were seriously threatened by failing to meet the production goals.

A few days later, the gleaming manager's vehicle jolted by the unpaved and potholed streets of the village outskirts, far from the city center, trying to find the place where he was expected to meet with the man who should help him solve his problem.

The barefoot children who were playing in the street, amid the fetid sewage that flowed next to the curb, stopped to give way and looked curious since that type of vehicle was not common in that poor place.

He had to stop several times to ask for directions before he finally got to the address he had written down on a piece of paper.

In front of the small wooden gate of the small and poorly plastered house, Antonio stopped and looked at the house surroundings. He wondered how he would be ridiculed if his colleagues knew where he was and, mainly, the reason for his visit to the site.

He wanted to get in the car and leave as fast as possible when a badly dressed and smiling black girl opened the door of the house and welcomed:

---

[1] Note: Preto velho (old black man): In Umbanda, an Afro-Brazilian religion, those are spirits of old slaves who died enslaved and have knowledge of divine magic. They are wise, peaceful, and kind spirits who know all about suffering, compassion, forgiveness and hope. They also prescribe herbal remedies.

"Good morning mister! Are you looking for something?"

"Oh yeah… I'm looking for an old man named Father Jau I guess. Is this where he lives?"

"Yes it is; he is my grandfather. Come on in, the gate is not locked."

Antonio hesitated, looked around once more, and finally entered the house where a black man with his hair all white, appearing old, waited standing in the middle of the small living room.

With a short greeting, Antonio was invited to sit on the small couch while the old man settled onto a wooden bench.

Antonio was anxious to get it over with and was direct to the point that brought him to that place:

"I think you know the reason why I came here. Don't you?"

"No, I do not know, sir," said the old man. "Mariano came to me saying that an important man from the company he works for wanted to talk to me about some problems he was facing and needed help, just that. Please, can you explain me what are your problems?"

The old man moved the wooden bench to the couch, lit a small clay pipe, and said quietly:

"And you can lay down your worries, it's just us at home. My granddaughter has just left for school."

Antonio reported in detail the problems that he was facing in the company and Mariano's suggestion about the origin of these problems in something supernatural, in his words a 'job' or a 'done thing' and concluded by saying what he heard about the assigned powers to Father Jau to 'undo' all the evil that had been done.

At the end of the narrative, a few seconds of silence took place. The black man gave a whiff of the small pipe and stared at the visitor looking deeply in his eyes, choosing the best words to answer:

"You know good sir, I'm just a poor guy who lives in these conditions you are seeing. Despite my poverty, I have a lot of faith and try to help these poor wretches who surround me, forgotten by governments and society, as best I can; at times saying prayers, at times advising or helping in the nursery we got here in the neighborhood, that's it."

And after another puff of his pipe, he continued:

"Sorry to disappoint you but I do not have the powers that some people like Mariano attribute to me. In addition, I do not believe these problems that you told me would be solved by rituals or prayers. They're something for skilled men like you to solve. I see in your eyes that you are desperate and the desperation, most of the time, leads us to wrong paths that not only do not lead us to the solution but make matters worse. Try to calm down and reason and do not let your work end with your life and the lives of your family because there is nothing more important in this world than this. Sometimes we are in the wrong place trying to do something we don't like or which we have no competence for. We have much to learn from nature, just observing the pigs, for instance: they were born with no wings, so we can't expect them to fly, it's against nature!"

Those words sounded like a mercy shot at the already shaky self-esteem of the poor Antonio. He didn't want to hear any more, just wanted to get out of that place as fast as possible and go to the company where he worked and deliver his letter of resignation.

Sad story, sad ending.

One of the reasons why I wrote this book, in addition to those already mentioned, was to show with real stories and facts that industrial maintenance doesn't need to be like the one in this story.

Many people don't believe this, but in fact, the maintenance staff working as 'firefighters' is not the right way to act and can be done differently; the same way it's possible to write and tell stories about industrial maintenance with happy endings.

How is this possible?

This will be explained in the best possible way by narrating my experiences in the form of stories, which will keep the reader interested and learn all about maintenance.

# 2 Concepts and Definitions
## *"Let's Call a Spade a Spade, Not a Gardening Tool"*

Before we proceed, it is advisable to review some concepts and definitions that will be used throughout this book.

I know that discussing concepts and definitions can be boring, and many readers would be tempted to skip this chapter, but I kindly ask those who have this intention, don't do it because it is very important what we are going to discuss.

It is common to find people, who work in the maintenance area, with divergent concepts about the same term or indicator and this situation, obviously, causes a lot of confusion, making it difficult to understand each other, and the problem may be bigger than it appears.

Let's start by defining the theme of this book: maintenance.

Maintenance: All actions necessary for retaining an item in or restoring it to a specified condition (United States Military, 1998).

If we analyze carefully the above definition, we will notice that the mentioned actions have two distinct objectives:

1. Keep the item in the state in which it can perform the required function. In this case, the word "maintain" has the sense of preserving, sustaining or preventing this state from being altered and, consequently, the item can no longer perform the function.
2. Restore the item to the state in which it can perform the required function. Now, the actions aim to restore the state in which the item can perform the function.

From this interpretation, we can conclude that there are, basically, two types of maintenance: preventive (1) and corrective (2).

In Figure 2.1, we see the different types of maintenance, which can be understood as variations of preventive maintenance. They will be described in sequence.

Design-Out Maintenance: It is the maintenance type whereby changes or modifications are done to the equipment aiming to eliminate the causes of failures, simplify or make safer the maintenance tasks, such as inspection or parts replacement, and improve the machine performance.

This type of maintenance is used to adapt the equipment to the operational requirements, especially when the equipment has frequent failures or high repair

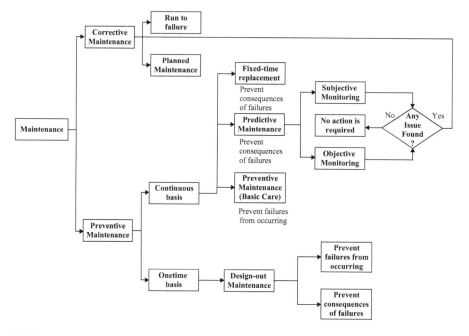

**FIGURE 2.1**  The different types of maintenance.

time due to conditions inherent to the equipment itself, usually because of poor initial design. The following actions are examples to increase the component/equipment life: improve access to lubrication points; add protective guard to protect electric motors, whose degree of protection is not suitable for water jets, against water penetration; review materials to prevent accelerated wear or corrosion; strengthen equipment foundation; and add protection to spiral hoses on mobile equipment to prevent abrasion.

The following actions are examples to ease maintenance interventions: improve the design of equipment protections to ease inspections, improve equipment access, install adjustment bolts, improve identification of inspection points and equipment adjustments, and install lifting points where necessary to facilitate the removal of equipment.

When deciding to promote some modification in the original design of equipment or its components, it is extremely important to verify that this modification will not have any undesired effect on its operation, especially, that no negative impact on safety will occur.

On changes being made without considering all the implications, it is worth remembering a terrible disaster well documented in world media that occurred in 1974 at a caprolactam (petroleum-derived substance, used as a monomer to make synthetic fibers, plastics, etc.) production plant in the city of Flixborough, England, where an explosion followed by a large fire completely destroyed the factory premises of Nypro (UK). Twenty-eight workers were killed and a further 36 suffered injuries.

Offsite consequences resulted in 53 reported injuries. Property in the surrounding area was damaged to a varying degree.

In summary, the tragedy originated in reactor No.5, which was leaking cyclohexane (a flammable liquid used for the industrial production of caprolactam), due to a vertical crack. The decision was taken to remove it and install a bypass assembly to connect reactors No.4 and No.6, so that the plant could continue production. In order to return to production as soon as possible, the plant modification occurred without a full assessment of the potential consequences; only limited calculations were undertaken on the integrity of the bypass line.

After 2 months of operation, a rupture occurred in the 20 inches bypass. This failure was attributed to the poorly designed project, since the precarious structure with scaffolding, installed to support the pipe, did not support its movement, due to the vibration that the piping was subjected to during the operation. This resulted in the escape of a large quantity of cyclohexane, forming a flammable mixture and subsequently found a source of ignition. There was a massive vapor cloud explosion which caused extensive damage and started numerous fires on the site (Health and Safety Executive, 1975).

Another very important point to consider when promoting a change in equipment or its components is the return on investment (ROI). In other words, the money to be invested in the modification should be justified by the benefits obtained with the modification.

Preventive Maintenance (ongoing): All actions performed to retain an item in specified condition by providing systematic inspection, detection and prevention of incipient failures (United States Military, 1998). Some examples are cleaning, lubrication, adjustment, alignment and balancing.

Predictive Maintenance: It is a set of activities to diagnose the physical condition of equipment (level of deterioration) during its operation to detect incipient failures of equipment, to determine the maintenance actions required and to restore equipment to its operable condition after detection of an incipient failure condition. Examples are vibration analysis, oil analysis, thermography, motor circuit analysis and ultrasonic testing.

To better understand the concept of Predictive Maintenance, also known as Condition-Based Maintenance, we will present the concept of the P-F or P-F Interval.

Let us examine Figure 2.2 that illustrates the process of condition deterioration of a given item.

This curve is called P-F because it illustrates how the failure starts and continues to deteriorate to the point at which it can be detected (Point P) and then, if it is not detected and corrected, continues to deteriorate, usually at an accelerated rate, until it reaches the point called F (functional failure), which is a point where the equipment is not able to continue performing a given function expected of it.

If the equipment remains in operation, despite the unsatisfactory condition, it will continue to degrade until it reaches the point where the total failure occurs, also called a catastrophic failure.

For the sake of illustration, let's imagine a hydraulic pump that needs to fill a process with a certain liquid at a given flow. In this case, the monitored condition

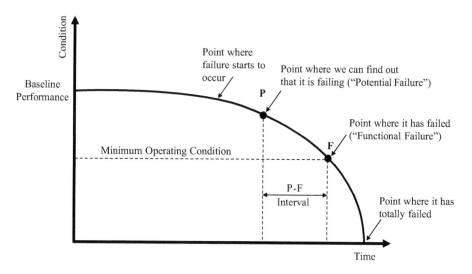

**FIGURE 2.2**    The P-F interval.

is the pump flow rate. Let us also assume that the liquid to be pumped has solids in suspension, and this will deteriorate the pump components and, by consequence, the liquid flow rate will decrease with time.

With increasing wear, and the progressive reduction of flow, the flow curve (P-F) will reach the point P (potential failure). At this moment, the reduction of the flow begins to be perceived in the process where the liquid is used. If nothing is done, the wear continues to increase and the flow decreases until it reaches the minimum operating condition; in other words, the minimum flow for the process to operate, the F point (functional failure).

Suppose the pump continues to operate, the wear will reach the point where any liquid will be pumped and we will have the catastrophic failure of this pump.

In addition to the above example, we have several other practical examples of potential failures that can be monitored. For example, excessive vibration in rotating equipment bearings indicating imminent bearing failure, increase of temperature ("hot spot") in electrical connections indicating possible contact problems and metal particles in gearbox lubricating oil indicating probable gear failure.

And what is the significance of the time interval between point P and point F in the P-F curve?

This time interval between the point at which failure can be identified and the functional failure is the available time for corrective actions to be taken to prevent functional failure from occurring or to avoid its consequences.

The method to be employed for potential failure detection will depend on the symptom and available methods. As we can see in Figure 2.1, Predictive Maintenance is subdivided into two different groups of methods to identify possible failures: Predictive Maintenance using subjective monitoring methods and Predictive Maintenance using objective monitoring methods that will be described below.

Predictive Maintenance—Subjective Monitoring: In this case, the monitoring is done using the human senses (look, listen, feel and smell).

The use of the human senses is one of the most basic and intuitive techniques of monitoring, and it can be very cost-effective because it requires practically no investment, except the necessary training. Another advantage of this type of monitoring is that the human being can detect a wide variety of failure conditions and is able to exercise judgement about the severity of the potential failure, providing the most appropriate action to address the problem.

However, before adopting this type of monitoring, some disadvantages should be considered. Usually, when the human sense is able to detect most of the abnormal conditions, the process of deterioration is already at a quite far advanced stage, and this means that there is not much time to take any action before the functional failure occurs (see Figure 2.3).

As the name itself says, this is a subjective process, so it is difficult to establish precise inspection criteria, and the observations will vary according to the experience and sensitivity of who is conducting the inspection. Moreover, the perception of a condition can vary from individual to individual. In summary, this type of monitoring cannot be expected to be accurate or repeatable.

Predictive Maintenance—Objective Monitoring: This type of monitoring requires the use of advanced technologies to determine equipment condition, and potentially predict failure. There are two options for monitoring:

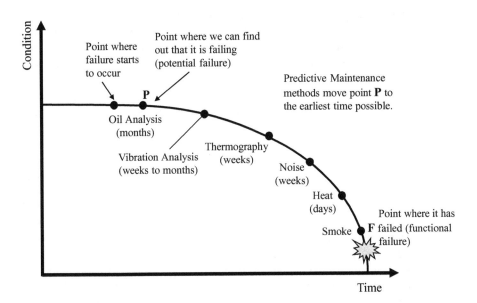

**FIGURE 2.3** Different P-F interval according to the predictive method. (Based on Blann, 2013.)

1. Off-Line Monitoring: It is performed at regular intervals through inspections and data collection in equipment. This monitoring must comply with a previously established predictive maintenance plan, and data collection is done according to preestablished routines.
2. Continuous Monitoring: It consists of the use of sensors permanently installed on the equipment, and their signals are captured on devices that read the parameters and, in some cases, may trigger an alarm or even turn the equipment off if the measured value exceeds a certain preestablished value in its configuration.

Currently, with the advancement of technology, especially the development of increasingly smaller and cheaper sensors, we have a dissemination of the use of predictive maintenance techniques in industry in general.

In Figure 2.3, we have a comparison between the various predictive methods (subjective and objective) and the required time for each of them to detect a mechanical problem in a gearbox.

Fixed-Term Replacement: This type of maintenance, which was very common in the past, provides for scheduled overhaul or component replacement, often based on the equipment manufacturer's recommendation or experience of local maintenance personnel with the type of component in question or still based on analysis of the history of maintenance of a certain equipment and its components. Some examples of this type of maintenance are automotive oil changes every 3,000 miles or the replacement of the carbon brushes of the DC motor every 6,000 h of operation.

When adopting this type of maintenance, it is important to keep in mind that, according to Idcon, 2019, only 10%–15% of all types of components have a predictable failure rate. In other words, there is a great chance of disposing of a component that is still far from the end of its useful life and, in this way, the maintenance cost is unnecessarily increased.

In summary, the cost of inspection or monitoring should be compared with the cost of replacing components at a predetermined frequency regardless of its condition.

Another hypothesis in which substitution may be feasible is at scheduled shutdowns in continuous processes where replacement of parts is made by opportunity.

Run to Failure: This is the type of maintenance where the repair is only performed when equipment has failed. Contrary to what may be imagined at first glance, this is a deliberate choice, considering the equipment failure mode and that the failure will not entail risks to the safety or operation, there will be no significant impact on the total maintenance costs and failure should be obvious, in other words, the operation personnel can easily identify that the equipment or component has failed.

This is the recommended maintenance strategy when the probability of failure of the equipment or component is constant, that is, it does not change over time, thus not allowing to determine any type of maintenance task to avoid failure or mitigate its consequences. It is noteworthy that according to Ivara, 2007, 33% of existing failure modes can be treated as "Run to Failure" either because there is no applicable maintenance technique or because prevention is not economically feasible.

As example of a component that will have "Run to Failure" maintenance type as recommended strategy is the lightbulb on the factory roof; its failure will not cause a threat to life, nor have a significant impact on company profits.

However, equipment or components whose maintenance strategy is adopted "Run to Failure" must have an established Contingency Plan that should be adopted as soon as the failure occurs to reestablish the operation in the shortest time possible, minimizing the impact of failure. Important to mention that without a Contingency Plan, the "Run to Failure" strategy is just emergency maintenance.

The suggested Contingency Plan should contain the following information about who will be responsible for the work, the estimated repair or replacement time, the internal or external parts, tools needed, the parts in catalogs, supplier data (name, address and telephone number), special safety precautions, etc.

Detective Maintenance: It refers to the tasks designed to determine whether a failure has already occurred. This type of maintenance is applied to hidden failures, those that aren't evident to operators when they occur. It is applied mainly to protective devices and alarms because the only way to find out if they will work when needed is to test them.

Some examples of Detective Maintenance or detective tasks are Pressure Relief Valves (PRV) tests, smoke or fire detection system tests and testing of relays protecting electrical equipment.

Continuing the definitions, let's discuss another extremely important concept that does not always have a common understanding: reliability.

Reliability: (1) The duration or probability of failure-free performance under stated conditions. (2) The probability that an item can perform its intended function for a specified interval under stated conditions (United States Military, 1998).

Let's discuss this definition for a better understanding; the abovementioned definition addresses the following important aspects:

- It talks about probability, that is, chances of survival. Observe that there is no absolute guarantee of the survival of the item (component, equipment or system).
- Performance without failures, that is, reliability ends when failure occurs.
- Note that the operation is defined for a certain period. For example, the expected operating time of a rocket carrying a spacecraft into space is quite different from the running time of an aircraft making its commercial route.
- The correct operation is a key element for survival.
- We must be fully aware of the specified environmental conditions; survival in one environment may represent failure in another.

Here are some other important definitions related to reliability:

Failure: The termination of an item's ability to perform its required function.

Failure rate ($\lambda$): The total number of failures within an item population divided by the total number of life units expended by that population during a particular measurement interval under stated conditions (United States Military, 1998).

Mean time between failures (MTBF): A basic measure of reliability for repairable items: the mean number of life units during which all parts of the item perform within their specified limits, during a particular measurement interval under stated conditions. It is calculated by dividing the total number of failures by operating time. It is also expressed as one over the failure rate. $MTBF = 1/\lambda$

The term MTBF is used for repairable systems, while mean time to failure (MTTF) denotes the expected time to failure for a non-repairable system.

Mean time to repair (MTTR): A basic measure of maintainability: The sum of corrective maintenance times at any specific level of repair, divided by the total number of failures within an item repaired at that level, during a particular interval under stated conditions (Figure 2.4).

Availability, Operational (Ao): The percentage of time that a system or group of systems within a unit are operationally capable of performing an assigned mission and can be expressed as Uptime/ (Uptime+Downtime). It includes logistics time, ready time, and waiting or administrative downtime, and both preventive and corrective maintenance downtime. This value is equal to the MTBF divided by the MTBF plus the Mean Downtime (MDT). This measure extends the definition of availability to elements controlled by the logisticians and mission planners, such as quantity and proximity of spares to the hardware item (NASA, 1997).

Availability, Inherent (Ai): The probability that the system is operating satisfactorily at any point in time when used under stated conditions, where the time considered is operating time and active repair time.

At this point, it is important to understand the difference between availability and reliability. Availability is a time-related indicator, that is, how long the equipment is available for operation while Reliability is related to the number of failures in each period of time.

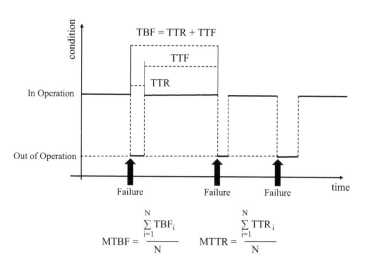

**FIGURE 2.4**  MTBF and MTTR.

In this way, the indicators can have totally different values; in other words, the same equipment can have high availability and low reliability.

Let us illustrate through a numerical example:

One piece of equipment failed three times in a year and each failure lasted 2 h. If the failure rate is constant and that the exponential distribution best describes its behavior over time, we will calculate the MTBF, failure rate, reliability and availability for this equipment on a 1-year mission or 8,760 h.

For exponential function, we have reliability expressed as follows:

$$R = e^{-\lambda t}$$

Failure rate : $\lambda$ = Number of failures/Mission time = 3/8760 = 0.000342 failures/h

$$MTBF = 1/0.000342 = 2924 \, h$$

The reliability will be : $R(8760) = e^{(0.000342 \times 8760)} = 0.0497\%$ or $4.97\%$

The availability will be $A = \left(8760 - (3 \times 2)\right)/8760 = 0.9993\%$ or $99.93\%$

One question I often hear when I present this exercise and its results in training sessions is: What is the practical meaning of these numbers?

With regard to reliability, we can say that the probability of not having any failure in the equipment in question, on a mission of 1 year or 8,760 h, is only 4.97%.

As for availability, we can say that the equipment will be 99.93% available to operate in the total time of 8,760 h.

The conclusion is that a reliable equipment or system has high availability, but an available equipment or system may or may not be very reliable.

With this in mind, I ask the reader the following question: Which is better; to have a production line with high reliability or high availability?

Any manager or production supervisor would respond immediately that he or she would like to have both high reliability and high availability; however, we should remind you that in real life, there is a high cost involved to have equipment with a very low probability of failure, such as an airplane or a nuclear reactor.

The most appropriate response should be that this condition depends on the characteristics of the operation; in general, a complex installation whose failure will seriously affect the safety of people or the environment or even have a difficult return to operation after failure, it will need to have high reliability.

However, in simple processes whose failure will not seriously affect safety, environment and the return to operation will be easy and fast, high availability is sufficient.

Here are some more indicators:

Maintainability: The ability of an item to be retained in, or restored to, specified condition, when maintenance is performed by personnel having specified skill levels, using prescribed procedures and resources, at each prescribed level of maintenance and repair.

Note: The term "maintainability" is used as a measure of maintainability performance.

A simple definition could be "the ease with which you can safely repair equipment in the shortest amount of time."

Qualitatively: Design features that give a machine the inherent condition of being maintained with a minimum of hours per person, without requiring high expertise, special tools or support equipment, and without offering security risks.

Quantitatively: The speed at which a machine can be restored to the operational condition after a failure or maintenance intervention.

With the concepts of reliability, availability and maintainability that we have just seen, let's look at Table 2.1.

From Table 2.1, as previously demonstrated, it can be noticed that if the reliability remains constant, even at a high value, the availability will not necessarily be high. As the time to repair increases, the availability decreases. In the numerical example presented, where availability and reliability were compared, it is noted that even a system with a low reliability could have a high availability if the time to repair is short.

An important conclusion that can be drawn from this table is that improved availability can be achieved either by improving reliability, reducing the number of failures or by improving maintainability, reducing the time to return the equipment in operation.

In general, if we compare the difficulty and the necessary resources to improve the reliability of an equipment comparing to what is necessary to improve the maintainability, we will conclude that it is relatively simpler and less expensive to improve maintainability, because much of what can be improved is maintenance responsibility.

For example, improvements in the training of maintenance personnel or in the spare parts management can bring significant reductions in equipment repair times.

In some cases, minor modifications to the equipment, as discussed when we introduced the Design-Out Maintenance, can greatly reduce repair times, improving maintainability and, consequently, availability.

**TABLE 2.1**

**Relationship between Reliability, Maintainability and Availability**

| Reliability | Maintainability | Availability |
| --- | --- | --- |
| Kept the same | Gets worse | Gets worse |
| Kept the same | Gets better | Gets better |
| Gets better | Keeps the same | Gets better |
| Gets worse | Keeps the same | Gets worse |

Based on Reliasoft, 2003.

Recalling the newly presented definition of reliability, I will present two cases to illustrate the importance of getting to know the design characteristics of equipment in order to avoid falling into the trap of trying to make the equipment operate in a condition for which it was not designed. Maintenance can only restore the equipment to the initial level of capability (or inherent reliability)—it cannot go beyond it.

## THE CASE OF PRODUCT CHANGE USING THE SAME PRODUCTION LINE

Some years ago, a company bought a factory from one of its competitors, which had decided to end its activities in the country.

After the acquisition, the purchasing company decided to launch a new product in the market and, to produce it, decided to adapt one of the lines of the newly acquired factory, thus avoiding new investments in equipment.

Although the acquired factory did not have all the detailed technical documentation of the line, the design engineers from the buyers concluded that the adaptation was relatively simple and inexpensive and it would not cause any major inconvenience.

The production line consisted of a series of synchronized conveyors, where the product was cooled until it was collected and the main change was the speed increase of the conveyors, which was one of the process requirements of the new product.

As the line speed increased, several problems appeared such as premature wear of components, overload in the cooling system and problems of synchronism between conveyors. In other words, it dramatically increased the occurrence of failures.

In summary, the line started to operate at a speed higher than the one specified in its original design, and the technical analysis was superficial and insufficient to detect all the changes that were necessary.

This type of situation, as incredible as it may seem, is common, and even more incredible is that the maintenance team has inherited the problem, being naturally appointed responsible for solving it.

At this point, I think the correct thing would be to admit the error and redo the line adaptation project, considering all the components and systems affected by the speed increase because, by that time, there were already real failure data to feed back to the project.

However, it was not what happened and there began the martyrdom of the maintenance team that started to repair the failures as they happened and, in a short time, everyone forgot that the line was modified and the maintenance team became the only one responsible for that problem, being marked incompetent because they could not eliminate all the repeating failures.

The maintenance team, on the other hand, did not try to find the root cause of the problems and only replaced the components when they failed to create an endless vicious circle.

In conclusion, we must never forget that simply observing the equipment design specifications, regarding the operating and environmental conditions, can avoid a series of disorders that will inevitably lead to failures and, consequently, losses.

Another important point that should be mentioned is that every company should have a formal system to manage the changes; to avoid situations such as the one just described; and to ensure safety of people, environment, facilities and equipment, broad consultation and approval of the modifications by all those involved, cost–benefit assessment of proposed modifications before implementation, and evaluation of the effectiveness of the implemented modifications.

The petrochemical industry reported serious accidents, such as the one reported earlier in this chapter, that could have been avoided if there was a truly effective Management of Change (MOC) program in place.

## THE CASE OF THE INGREDIENT WEIGHING SYSTEM

One company hired a new maintenance manager and soon after the first few days of work, his immediate superior, the industrial manager, challenged him to put into operation an automatic weighing system for ingredients used in the process that had never worked properly since its installation.

This system consisted of a series of silos, one for each type of ingredient used in the process. The ingredients had different forms; some in the form of small grains, others in powder form and others in pellets form, and each of these silos was interconnected to a single discharge cone, through appropriate feeders, and at the end of this cone, there was a load cell for weighing that mixture, which was dosed according to the preprogrammed formula on the operation panel and stored in plastic bags for later use in the production process.

The new maintenance manager tried unsuccessfully to collect data and information regarding the equipment to identify what kind of failures the equipment had. The most common answer he received from everyone about the equipment was simply that it did not work. No one was willing or did not know what kind of failure the equipment had, only said that it was faster and more accurate to weigh the ingredients in manual scales, completely abandoning the automatic system.

He also tried to get the production supervisor to perform some tests on the equipment so that he could observe the operation of the system and collect data that would lead to some conclusion, but the response of the production supervisor was always the same: "we have a production schedule to follow and we do not have operators and time to test; we've wasted a lot of time trying to make this 'gizmo' work, and besides, this is a production area not a laboratory" and ironically added: "I suggest you to paint this useless thing green and throw it on the grass to see if they forget this crap that only brought me more problems than I have!"

The existing documentation did not help either because it was a generic equipment manual with no information on design specifications, installation data or commissioning of the system.

In this way, "what cannot be cured must be endured" time has passed, other priorities have arisen and the system has remained forgotten in its place, static like the sphinx, waiting for someone, one day, to decipher it.

One day the maintenance manager was invited to a meeting where the weighing system issue would be discussed. In this meeting, the representative of the system manufacturer and the top management of his company would decide the future of the system.

As there were no data or even any structured information about the failures and their causes, the meeting seemed to be a discussion of philosophy or religion, or rather abstract, and that it would never come to any conclusion; if one side said that the equipment did not meet the specifications, the other side stated that the system functioned satisfactorily in several similar companies around the world.

The maintenance manager remained silent during the meeting as he couldn't add anything to that discussion and watched, bored, as the impasse formed, when the equipment manufacturer rep launched a challenge:

"I suggest you send an engineer to our headquarters in Europe, where the equipment is manufactured, to receive a free training on calibration of the system and, in addition, we will be more than happy to take him to visit some plants of our customers where similar equipment works satisfactorily for years and, in this way, he will be able to verify that the equipment works and, more than this, to learn how to calibrate it for such."

Those in charge of the company that bought the equipment accepted the offer, and the maintenance manager was assigned at the same meeting to this task.

Thus, with the process specifications and their tolerances in the suitcase, he landed in the middle of European winter, in the country where the equipment was manufactured.

At the first meeting with the equipment manufacturer's engineers, who were responsible for the visit, he asked to review the equipment purchase specifications and he could find, with surprise and amazement, that the accuracy required by the process was superior to that provided by the system.

In other words, without changes in the process specifications, his mission had already been completed; doomed to fail even before its beginning.

# 3 How to Change the Situation (Without Having to Appeal to Black Magic)

Some years ago, I assessed the maintenance processes of an electronics industry located in an Eastern European country and the situation that I found there was very similar to what was described in the story of poor Antonio; the availability of production lines was being seriously affected by frequent equipment breakdowns.

Maintenance staff was overwhelmed and totally involved only in corrective maintenance, in other words, "just working to put out fires," and the plant operations team was increasingly dissatisfied because they were not able to meet production schedules. The vicious circle was established; equipment broke, the production schedule was not met, the plant manager complained, operations team blamed maintenance team, and the maintenance team worked even more without obtaining good results from their actions.

A few months after the conclusion of the assessment and the issuance of the report with my observations and recommendations, I was called again to assist them in the difficult task of reversing the chaotic situation described earlier.

At the very first meeting I had with the maintenance manager and his team, I was able to note that they were all anxious for me to say that I had in my baggage the solution to their problems, however unrealistic this may seem, because for them the overall situation was unsustainable. Maintenance was the final element of the chain, that is, the pressure of the company's management because the production plan was not being accomplished would fall on the production team, and this amplified and passed on to the maintenance team.

Not surprisingly, I received the following question from the maintenance manager: "You came here, accurately assessed our maintenance processes and indicators, made several recommendations, but for us it is not clear what should be done to change this situation. Would you give us the 'magic bullet'?"

I thought about making a joke to relax, as I had done in a similar situation, when I answered those who asked me a similar question: "My name begins with J but it is Jose and not Jesus, and I do not do miracles" but, I reflected and I decided to be quite objective and didactic.

I went to a whiteboard in the conference room where we were and drew a horizontal line dividing the board more or less in the middle; at the upper side of the board, I wrote "proactive," and at the lower side of the board, I wrote "reactive" and began to explain my suggestion.

I told them that they were working only in the reactive mode, that is, simply react-
ing to failures after their occurrence, and to overcome this problem, the first step
would be to discover the root cause of failures to prevent their recurrence. Therefore,
for every new failure the question should be: "Why did it failed?" And then there are
two options:

1. The cause of failure can be identified: In this case, we must have an
   appropriate format to accurately describe the failures, such as failure codes,
   for example, and group the failures, facilitating the subsequent analysis by
   the reliability and maintenance engineers.
2. The cause of failure cannot be identified: For this second option necessarily, it is needed to apply the root cause analysis (RCA) methodology.

Addressing each equipment breakdown this way, we begin to understand what
is actually contributing to the poor production technical availability. However,
this is not enough to achieve our goals of reducing failures because we need to
be proactive, in other words, the failures need to be prevented from occurring or
the impact of occurrence must be minimized, and for this we have the preventive
maintenance plans.

These plans must originate from critical equipment analysis using methodologies
such as Reliability-Centered Maintenance (RCM) or Failure Mode and Effect
Analysis (FMEA). Although unnecessary, I made a point of emphasizing that the
mentioned methodologies should consider the recommendations of the equipment
manufacturers, the history of the failures and the existing maintenance plans.

In short, we must work in both modes, reactive and proactive, simultaneously.

Over time, for every failure that impacts the plant performance, we must ask why
the existing plans were not able to avoid it or minimize its consequences and the
answer of this question will serve as feedback to the system, promoting the review
of the existing plans.

It is established, then, a process of continuous improvement; initially, we will
have a stabilization and then a significant reduction in failures.

The plant performance should be monitored through different indicators
(availability, Mean Time between Failures, etc.) to verify how the maintenance
activities are influencing them and, consequently, to assess the success of this initiative.

I also highlighted that some basic conditions would be essential to be successful with this plan: The top management of the company must commit and
sponsor this initiative, training of all those involved in the need methodologies
and tools (RCA, RCM, etc.), effective participation of production along with
maintenance (team spirit), dedication and discipline of all involved persons and
clearly defined goals.

Then I took a picture of the whiteboard because I thought it would give a good
illustration to help me whenever I needed to address this subject again anywhere else.

After the explanation, I asked if what I had proposed made sense to them and if
anyone had any comments or doubts.

Initially, I was concerned about the silence that prevailed in the room but I decided
to allow to reflect and discuss between them what I had just exposed.

In the maintenance assessment that I had done previously, I had found that they had already implemented some of the methodologies that would be needed to change the situation such as RCA. In addition, they had many maintenance plans, mainly based on the recommendation of the equipment manufacturers, and therefore, I already expected the question that was made by one of the maintenance engineers, breaking the silence in the room:

> "But, we already have RCA, Maintenance Plans and we do most of the things you suggested. So, I do not understand why we will be able to change the situation and have good results from now on, if none of this has solved the problem to date. Could you please clarify?"

Although the answer was quite simple, I tried to choose the words and how to put them, avoiding to hurt anyone's feelings.

I replied that in a superficial analysis of the situation, the possible causes could be: problems in the implementation of the required methodologies (RCA, FMEA, etc.); lack of involvement of other involved parties, especially Production, in the problem-solving process; and, mainly, lack of a structured process where the causes of failures must, necessarily, feed back the revision of the maintenance plans to improve them.

I asked if my answer had satisfied them and nothing was said to me. There were only some discussions among the participants in the local language that I could not understand.

I was about to close the meeting when the maintenance manager informed me that the plant manager was expecting to have a presentation of the plan to improve the maintenance and, consequently, the production availability, before my departure.

That night, at the hotel, I created the process flow, as shown in Figure 3.1, from the picture taken from the whiteboard with my explanation, and I also prepared a draft of the maintenance improvement plan to present to the plant manager and his staff.

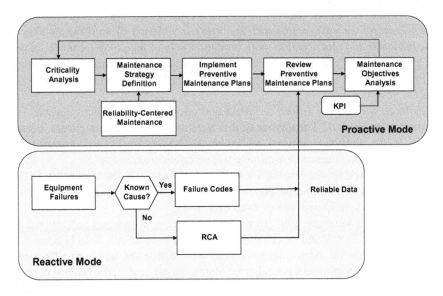

**FIGURE 3.1**   The maintenance process workflow.

In the meeting with the plant manager and his staff, I presented and explained in detail the diagram and, more than that, I asked for a vote of confidence and the participation of all involved in the process. I also suggested perseverance because the results would not be achieved immediately; in other words, it takes time for processes to be improved.

Sometimes the fact that you are a foreigner and having traveled a lot to get there gives you more respect and credibility than a local engineer, and I have used this fact to reinforce that although the plan was no "rocket science," I had full conviction that if it was implemented and followed with discipline, determination and, most important, perseverance, it would generate good results.

In the remaining days of my stay in the country, as requested, I provided training for several groups and participated in some RCA sessions to identify possible problems in the methodology application, helping them to improve the process.

For a number of reasons, including new professional activities, I could not follow the development and implementation of the improvement plan and its results.

So, I stayed for some time without receiving any news from that plant until one day when I was invited to participate in a project in Asia, I was surprised by one of these fortunate coincidences to know that the reliability engineer of that European plant, which had been the leader of the action plan that I described, would be a member of the project team.

In Asia, as soon as I met him, he told me that he was anxious to present me the results obtained from the implementation of the action plan that I had participated in.

I was really surprised how the results were achieved in a short time; the graphs in Figure 3.2 show the decrease in the number of failures of two production lines per week.

This impressive result was achieved due to the teamwork and dedication of all, as described by the proud European engineer who thanked me for the help and encouragement.

This is one of several real-life cases that I experienced where maintenance was able to reverse an unfavorable and troublesome situation of working only in the reactive mode and proceeded to work predominantly in the proactive mode, helping the company to reduce risks and meet its production targets, aligned with strategy.

From this point we will explain in detail each of the components of the presented process flow.

Failures: I invite the reader to reflect and answer the following question: Why do the equipment fail? The answer to this question is extremely important because it will be the starting point and will guide all actions so that we can effectively eliminate their occurrence or minimize their consequences.

Before answering the question asked, we will discuss the nature of failures and recall how understanding about failures and their relationship to the age of components and equipment has evolved over time.

According to Moubray, 1997, from the earliest days of industry until World War II, due to the predominantly mechanical nature and manual operation of equipment, they were understood to function reasonably well, with a low failure rate for a given time and then they entered a period of accelerated wear, that is, the failures occurred in direct proportion to the amount of time (or number of cycles) equipment spends in service.

The results – Weekly breakdown for X machine family

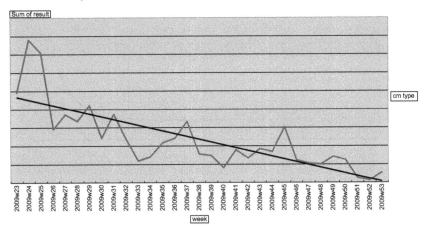

The results – Weekly breakdown for Y machine family

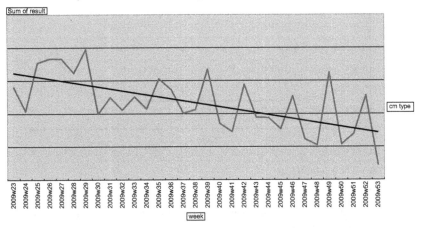

**FIGURE 3.2**  Graphs showing the reduction in the number of failures by equipment.

Briefly, the hypothesis was that the reliability of any equipment should be directly related to its operational age, suggesting that most items can be expected to reliably operate for a period of time, and then wear out as shown in Figure 3.3.

Over time, in the mid-1950s, there was an awareness of the high failure rate at the start-up of new equipment that gave rise to the famous "Bathtub Curve," so named due to its characteristic shape, which added in the curve the so-called "infant mortality" period (Figure 3.4).

In the 1960s, the aviation industry of the Unites States of America conducted extensive engineering studies on all of the aircraft in service to determine the source of failures and identified six different main failure patterns. The results of United Airline's exploration of the relationship between the failure rate or conditional probability of failure and some measure of operating age for aircraft hardware

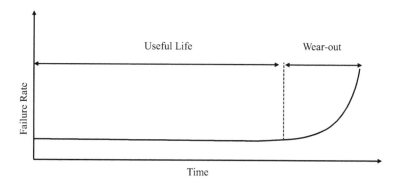

**FIGURE 3.3**   The wear-out curve.

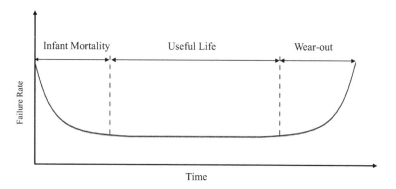

**FIGURE 3.4**   The Bathtub Curve.

are shown in Figure 3.5. The patterns were developed from a study of 139 aircraft components and equipment.

These results show both wear-out (age-related failures, curves A, B and C) and random failures (curves D, E and F).

The percentages found in Figure 3.5 showed that 4% of the items conform to the pattern A, 2% to B, 5% to C, 7% to D, 14% to E and finally 68% to the pattern F.

It is worth noting that these percentages are not necessarily exactly the same as those found in other types of industry, but there is no doubt that as the equipment becomes more complex, we will find more and more patterns E and F.

Let's, briefly, analyze each one of the above failure patterns:

- The failure patterns A and B were already mentioned before.
- Pattern C is characterized by a gradually increasing failure rate over the course of the equipment's life, without a defined wear-out zone. One possible cause of this failure pattern is fatigue.
- Pattern D shows low failure rate when the item is new, followed by a rapid increase to a constant level.

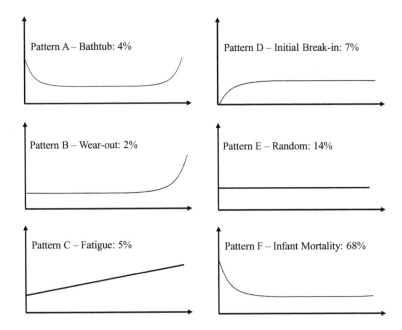

**FIGURE 3.5**    Six patterns of failure according to US aviation industry studies.

- Pattern E is known as the random pattern and is a constant level of random failures over the time.
- Pattern F is known as the infant mortality curve and shows a high initial failure rate followed by a random level of failures.

From the above patterns and studies, we have the following conclusions:

- Random failures predominate compared to age-related failures. Infant mortality (i.e. high initial probability of failure, decreasing with age) is common.
- The conditional probability of failure is never zero.
- Simple items tend to exhibit wear-out failures (patterns A, B and C), whereas complex items tend to exhibit random failures (patterns D, E and F).
- Patterns D, E and F should not be interpreted to mean that some items never degrade or wear out. These patterns simply show the life of some items came to an end before wear-out was evident, perhaps because they were removed for restoration, or were replaced with modified or upgraded items. Everything will eventually degrade with time, but some items degrade so slowly that wear-out is not a concern since the degradation will not adversely affect performance during the life of the equipment.

Given the above information, let's go back to the question that originated this thread: Why do the equipment fail?

According to reliability engineering analysis, we can assume that in general, except for special cases where there are redundancies, industrial equipment can be understood as a system resulting from several components in series and the reliability of this system would be the product of the reliability of the components of the system.

In this case, if a component fails, the whole equipment, as a consequence, will also fail.

In the example below, we will consider the calculation of reliability for 1,000 h of operation of a portable fan from the reliability of its individual components, considering that individual failures are independent events, that is, failure of one component does not interfere with others (refer to Figure 3.6).

Where the reliability of each individual component is: $R1 = 0.9995 - R2 = 0.9991 - R3 = 0.9915 - R4 = 0.9997 - R5 = 0.9998 - R6 = 0.9970 - R7 = 0.9515 - R8 = 0.9999$.

Then, the reliability of the fan, considering that all components are in series as mentioned before, will be:

$$R_{fan} = 0.9995 \times 0,9991 \times 0.9915 \times 0.9997 \times 0.9998 \times 0.9970 \times 0.91515 \times 0.9999$$

$$= 0.9028 \text{ or approximately } 90.3\%.$$

In other words, in 1,000 h of operation 90.3% of the fans would be in normal operation and 9.7% failing.

A first conclusion that we can take from the above is that if we want to have reliable equipment, with low likelihood of failures, we should have the components that make up this reliable equipment.

In summary, the equipment fails because its components fail and this answers the question we formulate at the beginning of this topic.

Now let's re-examine the Bathtub Curve (Figure 3.4) to understand the origins of the components failures. Currently, it is admitted that the Bathtub Curve is in reality the combination of two or more curves and its analysis is very important because, if we understand what influences each of its characteristic stages we will be able to understand the other failure patterns.

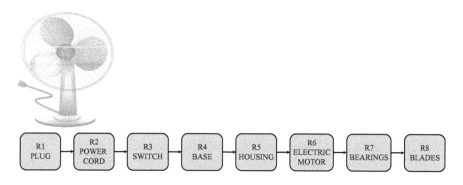

**FIGURE 3.6**  Portable fan reliability study.

In the curve above, we can see that a generic component presents three distinct periods, namely, infant mortality, useful life and wear-out, and then we will examine each of these periods.

Infant mortality: In this period, the premature failures occur and the failure rate is decreasing in relation to time and this could be due to maintenance induced failures or manufacturing defects in the components. There are numerous reasons why these failures occur, and we can mention the most common: design error, poor manufacturing process, operation error, out-of-specification materials, defective components, mounting error, incorrectly stored/handled components, contamination, commissioning issues, incorrect installation, overload in the first test, etc.

Has anyone heard the following phrase from the production staff: "Every time the maintenance team does preventive maintenance in this equipment, it does not operate anymore…."

This is a typical case of introduction of "infant mortality" in equipment when a maintenance task is performed, most often unnecessary.

For example, let's say that a given equipment has a component whose maintenance strategy is replacement with a specified time limit and, at the time of replacement of this component, an assembly error occurs that causes the recently installed component to fail. Consequently, this equipment that was operating normally ceases to operate by a failure caused by the maintenance itself.

What should be done to minimize the risk of components failing prematurely?

Below we have some suggestions of what can be done to reduce the risk of "infant mortality."

Implement the so-called "precision maintenance" with regard to installation and assembly: qualify the maintenance technicians and operators, establish formal commissioning and testing procedures, establish procedures for storing and handling components, and establish control of receiving components and maintenance materials.

Useful life (random failures): This period is characterized by the constant failure rate, and failures are of a random nature, but that does not mean that the failures cannot be predicted or mitigated. It means that overhauls, and replacements conducted at a specific frequency are not effective. Some examples of what may cause failures in this period are overload in operation, degradation due to the environment, human error during use or maintenance, improper application and unforeseen natural phenomena.

An example of this type of failure is as follows: The operator was washing his work area with a hose and, inadvertently, launched a jet of water directly inside an electric motor causing it to burn.

Some suggested actions to minimize the risks of these failures are review operation and maintenance procedures and evaluating improvements to avoid human error, establish or revise cleaning procedures and other basic care procedures, review need for damage protection, and evaluate feasibility of implementing condition monitoring.

Wear-out: This period is characterized by the increasing rate of failure as a function of time and the failures are usually related to the wear, erosion or corrosion and are often simple components which come into contact with the product.

The following are causes of wear: ageing, abrasion, erosion, corrosion or oxidation. Some examples are tires, mixer cylinders, pump rotors, etc.

It is suggested in case of wear failures either to observe the possibility of using other resistant materials or to check if the equipment operation is the most appropriate.

## FAILURE CODES

Who has ever needed to recover data on equipment failures, recorded in CMMS (computerized maintenance management system), or in any other media, where there is no structured coding system for failure description will know how difficult, or even, virtually impossible, is to do any statistical analysis since you can't group the data into failure classes.

To illustrate, let's assume that a particular conveyor has an electronic sensor that stops the material feeder when the conveyor is full of material. At any given time, the sensor fails and the feeder stops, even without excess of material. Most likely we would have the following descriptions in the maintenance work order, if it were completed by different people (e.g. operators, maintenance technicians, planners, supervisors): "the feeder stopped" or "lack of material in the conveyor" or "electrical defect in the feeder" or "full conveyor sensor failure," etc.

The process of repairing equipment, after a certain failure, involves a series of steps and people, starting with the equipment operator who detects the failure, going through the issuing and approval of the maintenance work order and the repair itself, carried out by the maintenance person until the end of the maintenance work order with the return of the equipment to operation.

Strictly speaking, the precise description of the failure of the equipment does not matter to most people directly involved in the above described process because, the person who will really need this information is the professional responsible to analyze the reasons of the equipment failures, a task normally attributed to maintenance and reliability engineers.

Therefore, people will have little or no motivation to improve on their own the information record, unless there is a structured, well defined, and mainly user-friendly failure coding system.

This is a point that deserves special attention from the maintenance person because it is necessary for clarity and awareness among operators and maintenance technicians for the correct completion of maintenance work orders. Therefore, they depend on the quality of the data that will be subsequently used by the maintenance and reliability engineers for the analysis of the failures, aiming at the improvement of maintenance plans.

It should be emphasized that another crucial point is the training of all users on the use of the codes. Follow the implementation process, checking the quality of the completed codes, correcting promptly the deviations and promoting meetings to report the results to all involved.

Before we continue with the "failure codes," I think it is important to understand the available means to record the maintenance interventions, and for this, we will briefly describe the CMMS. CMMS, in short, is a software package that is intended to document and manage all activities of a maintenance department among which we can highlight:

Register and update all the physical assets for which they are responsible, work order management, planning and scheduling of maintenance tasks, historical record of operations performed on assets, management of spare parts and maintenance materials, maintenance cost management, etc.

Currently, companies of all sizes use software to manage their maintenance. In general, these companies have someone assigned to administer the system and are granted access permissions to maintenance personnel, as required by the function. For example, the access of planners and maintenance programmers, usually, is different from the access granted to a maintenance mechanic who only records the information of the work orders he performs and, among this information, is the reason why the equipment stopped (cause) and what was done to repair it, in cases of emergency corrective maintenance.

Therefore, CMMS is the means available for managers and maintenance and reliability engineers to control, through the system-generated reports, the various maintenance performance indicators such as maintenance costs, productivity of staff, availability and reliability of equipment, and consequently, it is also in the CMMS that information on equipment failures should be obtained.

At this point, we will discuss another topic, important for the subject that we are addressing and that is the asset hierarchy structure, also known as equipment tree structure.

Asset hierarchy structure is one of the most basic elements of a CMMS. It is a logical structure to record machines and equipment in CMMS, allowing to have machines and equipment and the components that integrate them obeying a defined hierarchical rule, which also defines the dependency relationships between them.

This asset hierarchy is structured according to the physical location of the assets within the unit to which they belong and is assembled in descending order of the hierarchy. Each piece of equipment is identified by its position in the production process, and given a name and code.

According to the complexity of the process, this structure may have varying levels, but usually we have four levels: plant, system, equipment unit, equipment subunit and maintainable item/part.

In Figure 3.7, we have an example of an asset hierarchy of a Central Utilities, with five levels, where each layer of this hierarchy is detailed (gray rectangles):

First Level: Plant—Utilities
Second Level: System—Utility air
Third Level: Equipment unit—Compressors
Fourth Level: Subunit—Compressor unit
Fifth Level: Component/maintainable item—Casing, Rotor with impellers, Antisurge system, etc.

In summary, we can say that a well-structured asset hierarchy is an indispensable requirement to have the accurate information about the events and maintenance

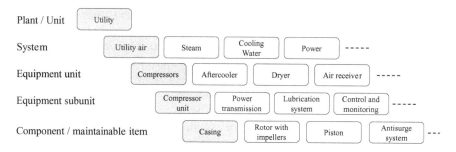

**FIGURE 3.7** Example of asset hierarchy.

interventions (downtime, employed hours, involved costs, etc.), exactly where they occurred for further analysis.

It is worth commenting that, on more than one occasion, I faced situations where it was necessary to redo the whole asset hierarchy structure by changing the way the CMMS was implemented because the originally defined structure, didn't allow the required information to be obtained for proper maintenance management.

Returning to the failure codes, we can, briefly, define them as a consistent way of documenting the key aspects of the failure event according to pre-defined categories. The codes can be numeric or alpha-numeric, most of the time, a logical abbreviation of their descriptions.

They are generally used directly in the CMMS, having the data entered by those who investigated and corrected the failure that originated the unplanned shutdown of the equipment. Currently, most CMMS allow users to register problems and subsequently supplement this information, with their probable cause and the adopted solution.

Now, let's deal with the other possibility, when the cause of the failure cannot be immediately identified and to do so, the RCA should be applied.

## ROOT CAUSE ANALYSIS

The root cause analysis or simply RCA, the acronym by which it is commonly known, is an indispensable methodology for the industrial maintenance to get out of the damaging reactive mode which we just described when we told Antonio's story in the Introduction of this book.

What prevents the successful implementation of RCA, even though companies are becoming increasingly aware of the benefits of RCA?

We must ask, "Why doesn't RCA work?"

Before continuing, for better understanding, it is appropriate to review some important definitions that are used in RCA.

Causal factors: All the factors that logically can affect results, including those that produce proven phenomenon.

Causes: The causal factors that are proven to cause or implied, directly or indirectly, of the phenomenon under analysis.

Root cause: The cause that, if corrected, would prevent recurrence of this and similar occurrences. The root cause does not apply to this occurrence only, but has generic implications to a broad group of possible occurrences, and it is the most fundamental aspect of the cause that can logically be identified and corrected. There may be a series of causes that can be identified, one leading to another. This series should be pursued until the fundamental, correctable cause has been identified (United States Department of Energy, 1992).

It should be noted that there is no single consensus definition of RCA in industrial, technical societies, corporations and companies. Each have their own definitions and it is rare to find two similar definitions. We opted for the following definition, which reflects the concepts discussed in this paper.

RCA can be defined as any evidence-driven systematic process that identifies the cause or causes of an undesirable event.

In the definition above of RCA, we can find one of the reasons why it doesn't produce the expected results in several places; the whole process should be driven by evidences.

It appears that many people do this routine bureaucratic process, organizing review meetings, preparing brainstorming sessions and applying the Five Whys technique without seeking any evidence. In other words, not investigating the "crime scene," not collecting samples, not interviewing witnesses and, at worst, not analyzing the real facts; they just cling to assumptions and then draw their conclusions.

In RCA training, it has proved relevant to establish an analogy between the RCA and the research methods of the famous detective Sherlock Holmes, a fictional character from the books of British author and physician Sir Arthur Conan Doyle. The first book about Conan Doyle's Sherlock Holmes, "A study in scarlet" originally edited and published by the magazine Beeton's Christmas Annual in 1887, is considered to be the first description in fiction of a scientific method of analyzing and solving problems.

The key points of the methodology used by the character to solve the most puzzling criminal cases, which are, in short, its hallmark and standard of conduct, are as follows:

- Sherlock Holmes had a passion for definite and exact knowledge (Doyle, 1887)—this implies that data should be collected to prove the hypothesis before determining a root cause.
- Believed that examining 1000 crimes would give information for solving the 1000-first crime (Doyle, 1887)—examine data from similar events because they will help to improve the process of analysis.
- He believed that the world was full of obvious things which nobody by any chance observed—don't accept the most immediate explanation without looking at all the information.

Accordingly, some quotes from Sherlock Holmes which applies to RCA:

- "It is a capital mistake to theorize before one has data. Insensibly one begins to twist facts to suit theories, instead of theories to suit facts" (Doyle, 1891).
- "Data! Data! Data!" he cried impatiently. "I can't make bricks without clay" (Doyle, 1902).
- "I never guess. It is a shocking habit—destructive to the logical faculty" (Doyle, 1890).

Throughout the explanation of the RCA process, the similarities between it and the methods of Holmes become more evident.

The RCA process that we have adopted basically consists of the following steps:

1. Define the problem
2. Collect data
3. Analyze the collected data
4. Identify effective solutions
5. Implement the solutions
6. Monitor the results

Step 1: Define the problem

Albert Einstein has said that if he had only 1 h to save the world, he would spend 55 min to define the problem and 5 min to solve it. The quote illustrates how important it is to define the problem in finding its solution. First, it is important to understand that any problem or undesirable event can be defined as the difference between the current situation and the goal (Eckert, 2005). A common practice in defining the problem, which ultimately hinder the subsequent analysis and solution, is that some people write a real novel describing the problem and, in most cases, end up defining not only a problem, but various problems in the same description.

We need to understand that different people or groups will have different views on the same problem (Eckert, 2005). One way to circumvent this difficulty and arrive at a consensus definition is to ask the following simple questions: What is the problem? When did this happen? Where did it happen? What goal has been impacted by the problem?

These questions must be answered in short sentences; one object and one defect.

Step 2: Collect data

At this stage, it is worth pointing out that, like Sherlock Holmes, we must collect and preserve the incident-related data in an appropriate manner for further analysis. To do this, we suggest that the data be grouped in the following categories:

- Situation: Examine the incident scene, identifying the situation. Check different dimensions, time and space. For example, location of incident, location of parts, control panel readings, time of incident, environmental conditions, size of spills and location of personnel.
- Witness: Identify people who need to be interviewed (witnesses). For example, technicians, operators, supervisors and engineers.
- Evidence: This mean something physical or tangible that need to be collected from the incident, such as samples of product, water, residue, grease, oil, failed components, tools, instrumentation and similar parts.
- Documentation: It is self-explanatory. For example, sketches, photos, design specifications, shift logs, process data sheets, plant line diagrams, procedures and manuals.
- Culture: It can be defined as the character and personality of the organization. It's what makes the organization unique and is the sum of its values, traditions, beliefs, interactions, behaviors and attitudes. Examples

are repetitious behavior, what people are used to (i.e. overriding "nuisance" alarms), condition of work environment (e.g. disarray, corroded parts in store), how people work (i.e. practice makes permanent, not perfect): "We've been doing it that way for years," "We kept reporting it to management," "I used to do that task but we removed it from our check list because we never found anything."

Step 3: Analyze the collected data

One of the methods used to identify possible causes of the problem is the causal tree analysis. The causal tree analysis starts by determining the so-called "main event," which is the problem or undesirable event being analyzed. This block is extremely important because it determines the rest of the sequence analysis. In sequence, it is necessary to determine what factors may contribute to the occurrence of the main event and the possible interrelations between them.

The relationship between the main event and its factors is the immediate cause–effect relationship.

The second level is the possible immediate causes of it. Thereafter, for each possible cause it should immediately be related to its possible causes, each immediate cause becomes an effect. And the diagram will be expanded to as many levels as needed, as shown below.

In a causal tree analysis, the main event is the incident itself and it is placed at the top or at left hand side as in the example below. The next step is to provide the causes for the top event, followed by the causes for those secondary causes, and continuing until the endpoints are reached. These endpoints are the possible root causes (Figure 3.8).

In determining the roots, to facilitate understanding of the event, the roots can be divided into the following categories: physical, human and organizational (or latent) roots.

The physical roots are the immediate consequences of the event; roots are tangible or damaged components, for example.

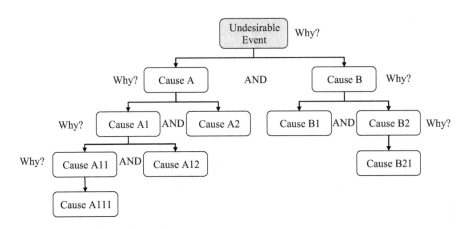

**FIGURE 3.8**   The causal tree analysis.

The human roots are the human actions that caused the physical roots or damage to components/materials, and finally the latent or organizational roots are the motivation for the action taken.

- Physical roots or the physical reasons why the parts failed.

  In order to identify the "physical" cause of failure, in some cases, the failure analysis should be performed. It is a detailed inspection of the damaged components to determine what was the mechanism or failure mode responsible for the failure. Failure analysis uses a wide array of methods, especially microscopy and spectroscopy. The information "how" the component failed is important data for determining the root cause. There are five mechanisms that lead to a component failure:

    Overload: The application of a single load (mechanical or electrical) leads the component to deform or fracture as the load is applied. Some possible causes are operation error or accident.

    Fatigue: Floating load over a relatively long period of time causes this type of failure and, in most cases, leaves clues. Some possible causes are thermally induced, mechanically induced, imbalance, misalignment, resonance and material.

    Corrosion-influenced failure: Corrosion substantially reduces the design strength of metals.

    Corrosion: The failure results is the wearing away of metals due to a chemical reaction. Some possible causes: wrong material, chemical process, environment, spills.

    Wear: Several mechanisms result in the loss of material by mechanical removal. Some possible causes: lubrication, contamination, misalignment and excessive loading.

- Human roots can be understood as human decision-making errors that will cause the roots of the physical event. They are errors of action or omission, which means, someone did something they should not have done or failed to do something they should do. Examples are memory (forgetting a task), selection (ordering wrong component, making wrong choice), discrimination (poor information), test or operation error ("knew" the rest of the procedure) and situational blindness (acceptance of problems).

  When the conclusion of an analysis is simply human error, there is a strong indication that the analysis was incomplete. Human error just says that something was not done correctly and that there were people involved. Human error is a general conclusion that does not allow any specific action to prevent recurrence of the problem. Once the specific cause of the problem was found, organizations choose disciplinary actions as the only alternative and keep thus a vicious circle.

  Organizations often blame employees for problems and seem to believe that this will set an example for all employees and discourage them to commit the same mistakes. In fact, the underlying message might be that "If you identify a problem or are involved with a problem which is preventing us from achieving our goals, it is better not to reveal because you can be punished."

- Organizational or latent roots can be understood as organizational systems which people use to make decisions. When systems are flawed, the decisions made from them will result in errors. Some examples of organizational roots are lack of employee engagement, management complacency, communication issues, task perceived as undesired, lack of procedures/ technical documentation, lack of formal training, incorrect incentive, use of incorrect or worn-out tools, incorrect priorities and lack of access to information.

Figure 3.9 describes the hierarchical order of root levels.

Step 4: Identify effective solutions

At this phase, the process identifies possible solutions for each individual cause found in the analysis mentioned above. It is important to verify that each solution prevents recurrence of the problem and does not create new problems. The ease of solution implementation and the required investment (cost/benefit analysis) should also be assessed.

Steps 5 and 6: Implementation of the solution and monitoring of the results

The whole process developed up to this point will be totally useless if the implementation of the solution does not take place. It is suggested that a complete plan must be prepared with all the planned actions with deadlines, resources and responsible persons for all actions.

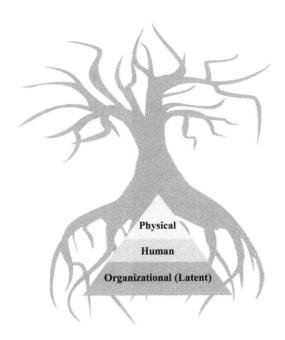

**FIGURE 3.9** The root cause levels.

The process of RCA aims at complete elimination of the problem preventing its recurrence. Recurrence at any time demonstrates that the process was ineffective for one of the possible causes: errors in determining the root cause, errors in the determination of actions to eliminate the root cause and errors in determining the parameters for monitoring the results.

### Resistance to the Implementation of RCA

To achieve success using RCA, you must be prepared to overcome the possible obstacles. Some of the arguments that people often use to justify their resistance are as follows (Latino, 2006): "RCA is just another name of witch hunt," "It is just waste of the time that we don't have. In fact, everybody already knows the causes," "It is just more forms to be filled up by us. In summary, more bureaucracy!", "RCA will never work here!", "Everybody knows that it is just flavor of the month," etc.

These arguments, however, are easily refuted. For example, those who think that RCA will take a lot of time, need to be reminded that if they do not have time for analysis, they will need to get more time and resources to handle the continual repetition of undesirable events.

Example of RCA methodology application—Case Description: In a tire manufacturing plant, there is an internal mixer equipped with two counter-rotating rotors in a large housing that shear the rubber charge along with the additives. The rotors were driven by a 2000 HP electrical motor/gearbox. In this case, the internal mixer of the tire plant had its automation system (PLC—programmable logic controller and software) replaced (upgraded) during a plant planned shutdown (holidays), after 5 years of continuous operation. As scheduled, it returned to operation on a Monday morning and operated continuously during that whole week until Saturday night, when it was supposed to stop for the weekend (the mentioned plant always stopped from Saturday night to Sunday night). At the moment when the equipment was stopped by the operator, an explosive sound was heard and some smoke comes out from the 2000 HP gearbox.

Problem: The internal mixer 2000 HP gearbox bearings were damaged.

Results from investigations that followed the event:

- It was a first-time failure for the equipment.
- The gearbox bearings were damaged due to lack of lubrication. The lubrication pump was stopped and the main motor continued operating.
- The operator used to shutdown the equipment by stopping the auxiliary equipment (hydraulic pump, fans, cooling water pump) instead of stopping the main motor. He claimed that he operated the equipment this way for more than 5 years.
- The engineer who made the PLC software conversion forgot to include the equipment safety interlocks.
- The equipment safety interlock wasn't tested during the equipment commissioning.

Causal tree analysis (Figure 3.10):

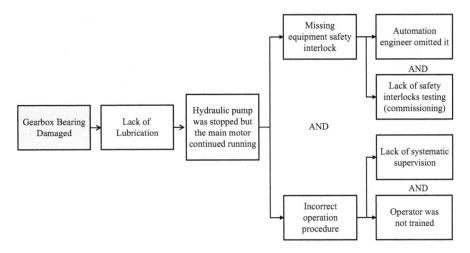

**FIGURE 3.10** The gearbox case causal tree analysis.

Root causes:

- Physical roots: The damaged bearings.
- Human roots: The operator didn't stop the equipment correctly, and the design engineer didn't include the interlock avoiding the motor to continue running if the lubrication pump is not.
- Organizational roots: Missing or inadequate qualification of operators, missing or inadequate commissioning of new/refurbished equipment, the area operations supervision was deficient.

Problem solution:

- The problem was solved by the replacement of damaged bearings and the inclusion of an equipment safety interlock to avoid the main internal mixer motor to continue running if the lubrication pump is not.
- The internal mixer operations procedure was revised and all operators were trained.

Extend the cause and fix:

To keep a failure from recurring is good, but the true value of RCA is to leverage the overall plant results. So, it was checked where else in the plant similar a problem could take place due to unknowledgeable and therefore wrong operation.

## THE CASE OF PRODUCT QUALITY VARIATION

Working as a maintenance manager, I once had a rather unpleasant experience of having practically my entire team involved in the overhaul of large and complex equipment, trying to find the root cause of a problem that was manifested in the product, affecting its quality (defective product) and, as a consequence, production losses.

It all started when we were forced to upgrade the process control system (supervisory software) of equipment, because the equipment manufacturer and developer of the software recommended updating the version of it for security issues and data integrity.

After the upgrade, the equipment ran normally during some weeks but then, all of a sudden, the product began to show variations in its consistency, that is, quality problems.

Due to the lack of RCA by process engineering, or whoever specializes in the product, as usually happens in most companies, the equipment immediately became responsible for the variation of the process.

Anyone who has experienced this kind of situation will know exactly what I am talking about. In the absence of a scientific methodology, lack of data to prove it left guesses and with each wanting to protect themselves, pushing the cause of the problem to another side, prevails the most authentic "teamwork," that is, mine against yours!

Those who attributed the product problems to the equipment, all except maintenance personnel, used the software update as an argument, forgetting or pretending to forget, that the equipment had generated products within the specifications of quality for a few weeks right after upgrading the software version.

Without much argument, because even the general manager was convinced the software was responsible for the problem, we had to return to the previous software version and, to the disappointment of the accusers, the problem in the product persisted.

Contrary to what one could imagine, those responsible for the process did not give up blaming the equipment and requiring the maintenance team to do, practically a general overhaul of the equipment; replacing control instruments, revision of the system of utilities, inspection in the mixing chamber and everything else that human creativity can imagine.

Until finally they have exhausted the possibilities of any problem in the equipment, and someone remembered that there could be problems with the raw material, the other variable in the system.

Bingo!!!

The raw material supplier, under pressure, admitted innocently that they had a "small problem" in their process and that some batches were shipped with deviation of parameters, and perhaps this deviation could cause the described problem.

The other day I was waiting for a doctor's appointment in a waiting room and, as the delay to be attended exceeded the time I had estimated, in the absence of better things to do, I decided to read one of those very old magazines, usually found in those places. While reading that magazine, one of these periodic variety magazines about history and science, which many people call useless culture, I stopped at an article about the Black Plague that hit Europe during the Middle Ages.

Researchers estimate that the pandemic decimated one third to half of the continent's population at the time, that is, something between 25 and 75 million people. In this scenario of despair and chaos, the population tried pointlessly to identify the cause of the plague and, what is worse, the lack of scientific resources and knowledge for research was compounded by ignorance and beliefs. In this way, the disease was attributed to the most varied factors among them the magazine highlighted: The divine wrath that punished humanity for their sins, or the black cats that were considered the incarnation of the demon on Earth.

Terrified, some of those who still survived killed the cats, hoping to get rid of the demon, while others self-flagellate trying to expunge their sins, thus avoiding contracting the disease. Years later, the immediate cause of the Black Plague or Bubonic Plague was identified as a bacterium called *Yersinia pestis*, transmitted to the human being through the fleas of the black rat (*Rattus rattus*) or other rodents. The rats found ideal breeding conditions in the urban agglomerations of the time, since hygiene in general and basic sanitation were virtually nonexistent—dirt and open sewage prevailed in cities and towns. The garbage was thrown in the street, the water was not treated and the contact with domestic animals was very close, with goats or sheep sleeping in the same beds as the people on winter nights.

Based on this information, eliminating cats or self-flagellation actually ended up contributing to the spread of the disease since cats are natural predators of rats and the self-flagellation weakened people, who were left with the body covered with wounds, making them much more vulnerable to pestilence.

As I read this article, I immediately associated with the episode I just reported, even with all the resources available for research, which were not even imaginable in the Middle Ages many people still insist on not using them, preferring to sidestep the problems instead of solving them.

# 4 Elementary! The Steps to Reach the Proactive Maintenance

Up to this point, we described the bottom side of the diagram, or just the reactive mode process; in other words, we described the actions that will be taken after the failure occurs.

In this chapter, we will describe the other side, the proactive mode (please refer to Figure 4.1): what we should do to avoid the occurrence of failures or, when this is not possible or economically feasible, to minimize their consequences.

Just as it was done for the reactive mode, we will briefly present each of the blocks that make up the proactive mode process flow and the methodologies associated with them.

Let's start with the first block on the right-hand side of the diagram, because first of all, the maintenance objectives need to be defined, which performance indicators will be used to monitor their performance and how they will be measured.

It is necessary to make sure that the maintenance objectives and their performance indicators are perfectly aligned with the company's business strategy because, as strange as it may seem, frequently, the maintenance objectives and strategy, as well as their performance measures are inconsistent with the organization's business strategy.

We'll describe the maintenance key performance indicators (KPI) later on in this same chapter.

Once the maintenance objectives have been defined, we will present the diagram from its left-hand side, starting with the criticality analysis.

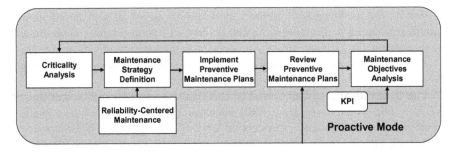

**FIGURE 4.1** Proactive maintenance process flow.

## THE CRITICALITY ANALYSIS

We assumed that we have a computerized maintenance management system (CMMS) installed and the asset hierarchy structure previously defined, as discussed in Chapter 3.

Initially, we need to understand why it is necessary to establish a classification of the plant equipment according to its criticality and, to that, I will use the same analogy that I employ when presenting training on this subject; let's imagine the maintenance of the family vehicle.

Let us assume that the family is on a restricted budget; in other words, the budget focuses on the family's immediate needs rather than its wants.

Assuming that the family will take a vacation trip and as precaution, the head of the family decides to do a review on the vehicle before taking the road.

After leaving the vehicle in an automotive workshop, it receives a budget detailing the services they recommend: repair the hydraulic brake circuit, replacement of internal reading lamps, repair small scratches on car paint and replacement of rubber mats.

Assuming that the money is sufficient to do just one of the services, which of the services would you do?

I believe that the obvious option will be to repair the brake system for the simple reason that a brake failure could endanger the car passengers as well as others on the road.

Even intuitively, a qualitative analysis of the criticality of the components of the vehicle that needed repair was made and the criterion adopted prioritized the safety of the people in detriment to the comfort and the appearance.

The equipment criticality analysis that we are going to present, analogous to the example of the family vehicle revision, has as main objective to classify the equipment in an industrial plant, with respect to its importance to the continuity of the operation, and the priority order is decided by the risk that the equipment failure represents.

There are several quantitative and qualitative techniques aimed at establishing a systematic way to decide which equipment should have priority within a maintenance management process.

In our case, we will present one of the widely used qualitative techniques which, on the one hand, apparently is not as accurate as a quantitative analysis and, on the other hand, has the advantage of being much more practical and faster because it basically depends on experience and knowledge of the evaluation participants.

Briefly, the methodology considers the likely impacts that a given equipment failure can cause to the industrial plant in terms of people safety, environment, quality, production, equipment utilization, estimated frequency of failures and cost.

The criticality analysis process outputs a single classification (A—high, B—medium, or C—low) for the evaluated equipment.

The criticality analysis should not be carried out by a single person working on his own. The principal objective of the assessment is to make use of the collective experience of maintenance, production and Health Safety and Environment personnel, and for this reason, it should normally be carried out using a small work group.

Whenever necessary, people should be brought in from other departments (Engineering, Stock Control or Finance) to give any specific advice.

The mentioned group jointly assesses the plant equipment criticality, and the group's first task is to detail the levels of factors according to Table 4.1, since the values in it are only for reference and will vary according to the reality of each plant.

The second step, starting from a list of all existing equipment in the production process, the team should classify them as the flowchart in Figure 4.2.

Once the plant equipment is assessed according to the defined criteria, the maintenance strategy needs to be set up for each equipment according to its criticality.

From experience, we recommend that the process be led by a production representative, and the maintenance and reliability engineering team must act as facilitator of the work.

## TABLE 4.1
## Criticality Evaluation Factors and Levels

| Evaluation Factor | Level | | |
| --- | --- | --- | --- |
| | 1 | 2 | 3 |
| Safety: Potential risks to the safety of people. | The equipment failure can cause fatality or serious accident. | The equipment failure may cause accident without injuries. | There are no consequences. |
| Environment: Potential risks for the environment. | The equipment failure causes serious effects on the environment. | Equipment failure affects the environment. | There are no consequences. |
| Quality: Effect of equipment failure on product quality. | The equipment failure affects the products quality. | Equipment failure makes the product quality vary and requires reprocessing. | There are no consequences. |
| Production: Effect of equipment failure on production. | The equipment failure causes total disruption of the production process. | The equipment failure causes an interruption of an important system or unit or reduces production. | There is spare equipment or it is more economical to repair the equipment after failure. |
| Working Time: Use of the equipment in the production process. | The equipment is used 24 h a day/7 days per week | The equipment is used only two shifts per day. | The equipment is used only occasionally. |
| Frequency of Failure: Quantity of failures in a determined period (failure rate) | High number of failures. For example, MTBF less than a month. | Occasional failures. For example, MTBF between 1 and 6 months. | Reduced number of failures. |
| Cost: Total cost for repair (material and labor). | High repair cost. For example, More than US $10,000.00 | The cost is between $5,000.00 and $10,000.00. | Low repair cost. For example, less than US $5,000.00 |

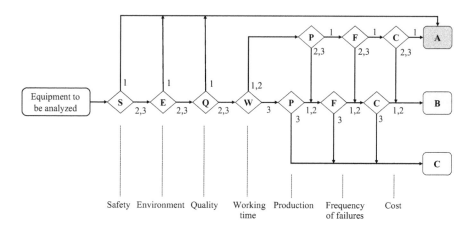

**FIGURE 4.2** Criticality analysis flowchart.

We also suggest that at each evaluation session, a register of attendance is kept for the purposes of registration and traceability of the process.

As stated above, the quality of the work will depend on the knowledge and experience of the members of the evaluation team, so it is indispensable that this requirement be followed.

The meetings should preferably be arranged by specialty (equipment family) to increase productivity.

Typically, this evaluation results approximately 25% of equipment classified as A, 35% of equipment classified as B and 40% of equipment classified as C.

It is very common, in this type of evaluation, to find participants, especially those linked to production, saying: "all our equipment is critical!", because they fear that equipment classified as C is abandoned by maintenance, which is not or it shouldn't be reality.

First of all, it is necessary to clarify to all participants that the more impartial and careful the analysis, the better the distribution of maintenance resources.

It is also worth pointing out that the described assessment reflects a momentary situation and the values of the evaluation factors may vary over time. For example, let's assume that a particular equipment was classified as C in function of the Working Time and production factors have been evaluated by the group as 3 because this equipment would have occasional use at the time, since the volume of production was low because of the marketing demand.

Supposing that any change in the market significantly increased the sales of the product, changing the production schedule, changing those mentioned factors to 1, consequently, the equipment should be classified as A, in this new scenario.

It is recommended that the criticality of the equipment be reviewed annually to verify that some equipment needs to have its criticality altered.

Finally, we warn that in the case of equipment regulated by standards and legislation, for example, pressure vessels, they should prevail over the described method.

## MAINTENANCE STRATEGY DEFINITION

To better understand this important step, we will use the same analogy that we discussed in The Criticality Analysis section when we present the equipment criticality analysis, that is, the maintenance of the family car.

Assuming that this vehicle was purchased secondhand, with a few years of use and no longer has the owner's manual, where it is possible to find the maintenance recommendations. In this way, the owner will have to define the maintenance strategy for his vehicle aiming to achieve two seemingly conflicting goals: maximum reliability at the lowest cost.

By maintenance strategy, we understand: The definition of the most suitable maintenance types to maximize equipment life and performance for the least cost, and to be able to define a successful maintenance management strategy, it is necessary to understand how equipment fails.

In the example of the family car, the following list illustrates some basic recommendations for routine maintenance:

- Regular checking of the vehicle's fluid levels, for example, engine lubricating oil, engine coolant, brake system fluid and power steering fluid.
- Change engine lubricating oil every 3,000 miles or 3 months, whichever comes first. Replace oil filter with every oil change.
- Keep tires inflated to recommended pressure. Check for cuts, bulges and excessive tread wear.
- Periodic checking of lights: All lights should be clean and working, including brake lights, turn signals and emergency flashers.
- Look for signs of oil seepage on shock absorbers.
- Inspect windshield wiper blades whenever you clean your windshield.
- Have the battery checked with every oil change. Cable ends should be free of corrosion.
- Look underneath for loose or broken exhaust clamps and supports. Check for holes in muffler or pipes.

In a concise and simplistic way, this would be the car's maintenance strategy.

It was elaborated on the basis of experience and similarity to what is recommended by car manufacturers in general but, what if we need to elaborate the maintenance strategy for an equipment that we do not know as well as an automobile?

Where should we start?

In order to develop a maintenance strategy for a particular equipment, similar to the proposed strategy for the family vehicle, we must first of all understand how the equipment fails.

As we have already discussed in Chapter 3, in the past, there was an intuitive belief that failures occurred in direct proportion to age/utilization of equipment, most likely by the fact that mechanical parts wear out with use. In this way, there was a hypothesis that the reliability of any equipment should be directly related to the operating age.

Another belief guiding maintenance managers at the time was that the probability of equipment failure could be statistically determined, and thus the predominant strategy was the "periodic overhaul" where the equipment was dismantled and its components could be replaced or restored, before the occurrence of failures.

The "periodic overhaul" was based on the principle that most equipment would have its life according to the "Bathtub Curve" (see Figure 4.3) previously discussed. Then, an intervention interval was established in such a way to prevent the equipment from reaching the wear-out phase, thus reducing the likelihood of failure.

The aforementioned research by the US civil aviation industry in the 1960s discovered that

- There was no strong correlation between age (time) and failure, which contradicted the basic premise of the "periodic overhaul," adopted without distinction for the vast majority of equipment.
- Statistical analysis did not demonstrate any improvement in the safety or reliability of the equipment when the overhaul intervals changed; it was believed that the reduction in the interval between interventions would improve reliability, but the results showed the exact opposite in many cases.
- Overhauls generated high direct and indirect costs for little or no benefit.
- As discussed in Chapter 3, most failure modes do not support the philosophy of time-based maintenance; they don't have wear-out characteristic.
- The change in the intervals between "periodic overhauls" was not normally based on analytical methods.
- Lose considerable component life with replacement based on time and not on condition.
- Overhauls most often reintroduce infant mortality failures.

From these findings, the industry created the so-called "Maintenance Steering Groups" (MSG) to re-evaluate the maintenance strategies of their aircraft. These groups were representatives of aircraft manufacturers, airlines and the US government. From these studies came the concepts of Reliability-Centered Maintenance (RCM).

The term "Reliability-Centered Maintenance" was first documented by F.S. Nowlan and H.F. Heap and published by the US Department of Defense in 1978.

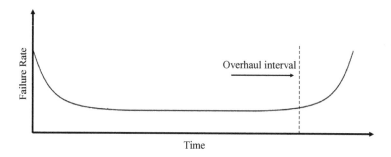

**FIGURE 4.3**   Bathtub Curve with overhaul interval.

Since then, the process has been redefined by different authors, altering some elements originally defined by the authors of the methodology.

In 1999, the Society of Automotive Engineers (SAE) published the SAE JA1011 standard called "Evaluation Criteria for Reliability-Centered Maintenance Process," which describes the minimum criteria necessary for any process to be called "Reliability-Centered Maintenance". However, this standard does not define any specific RCM process, and for this reason the SAE JA1012 standard was published in 2002 as the "RCM Guidebook: Building a Reliable Plant Maintenance Program." John Moubray's book, *Reliability-Centered Maintenance*, which was one of the references for the elaboration of the SAE JA1011 standard, surely is one of the main sources of reference for the study and implementation of RCM.

## RELIABILITY-CENTERED MAINTENANCE

RCM is the application of a structured method to establish the best maintenance strategy for a given system or equipment. It begins by identifying the functionality or performance required by the equipment in its operational context, identifies failure modes and likely causes, and then details the effects and consequences of the failure. This allows us to assess the criticality of failures and where we can identify significant consequences that affect safety, availability or cost. In summary, this methodology allows selecting the appropriate maintenance tasks directed to the identified failure modes.

It is worth mentioning that RCM focuses on the most important functions of the equipment or system and eliminates maintenance tasks that are not strictly necessary. In order to do this, the failure modes that affect the functions must be identified, and the importance of each functional failure must be determined from its failure modes to define the applicable and effective tasks to prevent these functional failures.

The RCM process is a structured approach that requires the analyst to justify maintenance requirements by answering seven questions:

1. What are the functions and associated performance standards of the asset in its present operational context?
2. In what ways does it fail to fulfill its functions?
3. What causes each functional failure?
4. What happens when each failure occurs?
5. In what way does each failure matter?
6. What can be done to predict or prevent each failure?
7. What should be done if a suitable proactive task cannot be found?

Our purpose is to briefly present the RCM and not delve into the subject with details on the application of the methodology. If the reader is interested in more information about the RCM, there are excellent publications that are listed in the bibliographic reference.

That said, we will briefly address each of the questions above, so that we can understand how the RCM analysis is done.

It is important to emphasize that this methodology requires time and dedication of several people; therefore, the analysis of all the equipment of an industrial plant is practically unfeasible. In this way, the analysis should be restricted to equipment considered critical (Class A, according to the criticality analysis previously done).

The first step of the RCM, which is the answer to the first question, is the definition of the functions of the equipment or system in its operational context with its respective performance standards required by the users.

Function can be defined as: Any action or operation which an item is intended to perform (United States Military, 2011).

We can divide the functions into primary and secondary functions. The main function of a physical item can be understood as the main reason for which the item was acquired, being important to emphasize that the RCM analysis is always initiated by the main functions. This category of functions covers details such as speed, production capacity, storage, transportation, product quality and customer service.

Secondary functions can be understood as what users expect from the physical item, in addition to fulfilling their primary functions. Users usually have expectations regarding: safety, control, containment, comfort, structural integrity, economy, protection, operating efficiency, meeting legal and environmental requirements, and even appearance.

In general, secondary functions are considered less important than the main functions, but they must be considered and analyzed rigorously because their loss can have serious consequences in certain situations. In some cases, the loss of a given secondary function may have a more serious consequence than the loss of primary function.

Regarding the operational context, mentioned in the first question, it describes the physical environment in which the equipment operates, the requirements of the process and the details of how it is used.

The operational context is one of the factors that most influences the equipment or system performance, that is, how close to its nominal capacity the equipment operates, what external environment it is in, how the environment can affect its performance and the existence of an installed spare unit that can be operated if the equipment fails.

This first step of the RCM process, that is, the definition of functions, according to Moubray, 1997, if carried out with the necessary rigor, should consume approximately one-third of the total RCM analysis time.

One indirect benefit that this step will bring to participants in the analysis is learning how effectively the equipment works.

Let's present an example of defining roles and performance standards to illustrate what was presented:

One pump to supply potable water should pump water from the water treatment reservoir (point A) to another reservoir at the point of consumption (point B):

- Primary function of the equipment (pump): supplying drinking water to the user with a flow between 800 liters/min and 1,000 liters/min.
- Secondary function: contains water, that is, no leaks.

The next question refers to how the equipment fails to meet user expectations, that is, functional failures. These can be defined as the inability of the equipment to fulfill a given function to a performance standard acceptable to the user.

In order to better describe functional failures, it is necessary to clearly understand the performance standards associated with the required functions of the equipment. For example, in the case of the drinking water pump, mentioned above, we can relate the following functional failures: The pump provides a flow rate of less than 800 liters/min, and/or the pump does not contain water and has leaks (Figure 4.4).

It should be noted that in RCM, functional failures may be partial failures where the equipment continues to operate at an unacceptable level of performance. In the example above, if the pump can continue to operate but, with a flow rate less than 800 liters/min, it will be considered a failure. This rationale reinforces the importance of defining both function and performance standards so that the analysis is well done and produces the expected results.

The performance standards are necessary to identify as the maintenance can only restore the asset to its initial level of capability (the designed capability or the built-in capability).

The third question requires us to try to identify all the events that may cause the functional failures of the equipment or system under analysis.

These events are known as "failure modes". Failure modes include those that have occurred in the equipment or that have occurred in identical equipment operating in the same operating context and failures that have not yet occurred but which can be considered plausible in the context in question.

Identifying the failure modes of a physical item is one of the most important steps in this methodology because, once each failure mode is identified, it is possible to verify its consequences and to plan actions to prevent failure or minimize its consequences.

Normal wear and tear are the causes of failure that are part of most lists of traditional failure modes. However, the list shall include failures caused by human error, either those caused by operation or maintenance personnel other than design failures, so that all causes of equipment failure are identified and adequately addressed.

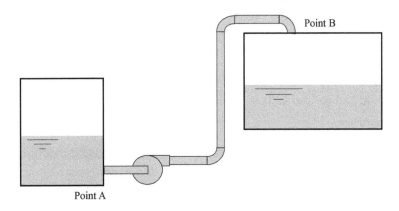

**FIGURE 4.4**   Water pump example.

It is worth mentioning that it is very important to distinguish a functional failure (failure state) from a failure mode (an event that could cause a failure state).

We will exemplify through an industrial water pumping system whose flow required by the process is 800 liters/min (Table 4.2)

The fourth question of the RCM process concerns the failure effects, that is, the description of what happens when each failure mode occurs.

These descriptions should include all information necessary to support the assessment of the failure consequences, such as:

- What is the evidence (if any) that the failure occurred?
- In what ways (if any) does it impact safety and the environment?
- In what ways (if any) does it impact production or operations?
- What is the physical damage (if any) caused by the failure?
- What should be done to repair the failure?

The effect described should be that which occurs if no specific task is being performed to anticipate, detect or prevent the failure.

It is important to note that failure effects are not the same as failure consequences. The failure effects describe "what happens," while the failure consequence is the description of how the loss of function matters.

The fifth question analyzes the consequences of failures, recognizing that the consequences of failures are more important than their technical characteristics.

In fact, the only reason to adopt the so-called "proactive maintenance" is not only to avoid failures but to avoid or reduce their consequences, when it is not technically or economically feasible to avoid them. The RCM process classifies these consequences into four groups as follows:

- Hidden failures, that is, they have no direct impact but they expose the organization to multiple failures with serious, often catastrophic, consequences, most of which are associated with non-fail-safe protective devices.

    As an example of a hidden failure, let us imagine a safety valve installed to protect a pressure vessel from excessive pressure buildup. Assuming that this valve, for any reason, is in a failure state so that it will not operate when the fluid pressure stored in the vessel exceeds the preset value.

---

**TABLE 4.2**

**Functional Failure and Failure Mode Correlation**

| Functional Failure | Failure Mode (Cause of Failure) |
|---|---|
| Unable to pump water | Seized bearing |
| | Pump rotor clogged by foreign object |
| | Electric motor burned |
| | Inlet valve locked in the "closed" position |
| Pumping water with flow rate less than 800 liters/min | Wear on pump rotor |
| | Suction line partially blocked |

Under normal operating conditions nothing will occur because, in principle, the nominal pressure of the vessel will be below the opening pressure value of the valve which, because it is in a failure state, will not act even if the pressure exceeds the maximum value that the installation supports. In short, the installation will be unprotected against excess pressure and no one will notice that the valve is failing.

In case of abnormal increase of fluid pressure, exceeding the maximum permissible limit by the system, as the valve that should protect the installation is in the state of failure, in the limit, a catastrophe may occur.

- Failures that have an impact on safety (if it causes injury or death) and the environment (if it violates any type of environmental standard or regulation).
- Consequences about the operation: The failure will have operational consequences if it affects production (production volume, product quality, customer service, or operating costs beyond the cost of repair).
- Non-operating consequences: Obvious failures fall into this category if they do not affect the other factors mentioned above, so it will only involve the direct cost of repair.

The sixth question relates to proactive tasks. They are defined as the tasks that are undertaken before a failure occurs in order to prevent the asset from failing. In other words, the definition of the applicable and effective tasks performed to predict, prevent or find failures.

This group of tasks encompasses what is traditionally known as preventive and predictive maintenance.

The last question is about identifying the tasks or activities that need to be performed when a suitable proactive task cannot be found which is both technically feasible and worth doing for any failure mode. These tasks or activities are called "default actions," and they are governed by the consequences of failure. RCM recognizes three major categories of default actions (maintenance types): detective maintenance also known as failure-finding tasks, design-out also known as redesign, and run to failure.

Recalling the definitions presented in Chapter 2, we propose the following types of maintenance:

1. Design-out
2. Preventive
3. Predictive
4. Fixed-term replacement
5. Run to failure
6. Detective maintenance

These defined types would fit into the RCM definition as follows:

- Proactive: (2), (3) and (4)
- Default actions: (1), (5) and (6)

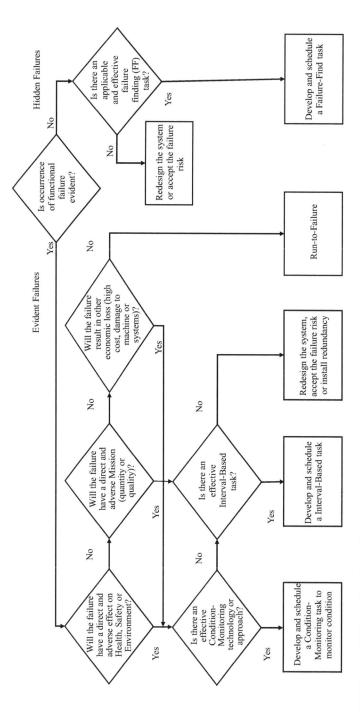

**FIGURE 4.5** RCM decision logic diagram.

After answering the seven RCM questions, it is necessary to integrate all the information obtained during the process and define the corresponding maintenance activities for each failure mode.

- Questions 1 through 5 can be answered through a Failure Mode and Effect Analysis (FMEA), which we will briefly summarize later.
- The remaining questions 6 and 7 have their answers through the RCM Decision Logic Diagram (Figure 4.5).

## FAILURE MODES AND EFFECTS ANALYSIS

FMEA can be defined as "analysis of a system and the working interrelationships of its elements to determine ways in which failures can occur (failure modes) and the effects of each potential failure on the system element in which it occurs, on other system elements, and on the mission" (NASA, 1997).

This analysis aims to anticipate known or potential failure modes and recommend corrective actions to eliminate or compensate for the effects of failures.

The FMEA should not be performed by a single person, which I have seen in some places where companies hired a consultant to do the job and left the consultant working alone rather than being the facilitator of a working group with people who actually knew the equipment and process under analysis.

The main objective of the FMEA is to make use of the collective experience of maintenance and production personnel to analyze causes of failures and propose solutions, and for this reason the FMEA should normally be carried out by a small working group with all those who can contribute to the quality of the analysis including, for example, a work group leader or facilitator, maintenance personnel, operators who know the equipment/system well, safety and environmental technicians, when necessary, consult people from other departments (Engineering, Planning and Programming, Materials and Finance).

An additional benefit of this method is that the joint participation of Operations and Maintenance teams in the FMEA generally increases everyone's knowledge of the equipment or system under analysis, and the proposed tasks will have been previously discussed and agreed by the group rather than unilateral Maintenance decisions.

### BASIC INFORMATION REQUIREMENTS

If the FMEA is performed in a company that has no maintenance history or a greenfield plant, the information needed for the analysis will be derived purely from the experience of the working group participants and the know-how of equipment manufacturers. Otherwise, the information required for the FMEA should come from the company's maintenance records.

Typically, the following information is required:

- Records of all breakdowns and stoppages of the production equipment under analysis for a period of not less than 12 months and, where possible, up to 36 months. These records must contain the damaged equipment,

subassembly or component, the tag number in the equipment tree, the symptom of each failure (as the problem manifested), primary and/or secondary causes, total downtime recorded including waiting time, if any, cost of the required labor, and cost of the required spare parts.
- Technical documentation, including plans, diagrams, manuals, layouts and flowcharts of the entire production process, preferably with all functional dependencies.
- Additional information on the operating conditions of the equipment (climatic conditions, typical conditions of use, environmental conditions, etc.)

## FMEA Step-by-Step

The working group, with the required maintenance data of the equipment or system under analysis, shall prepare the FMEA table which normally contain the following columns:

Column 1: Name of the assembly, subassembly, or component to be analyzed.

Column 2: Function—Remembering that function is understood as any action or operation in which an item is intended to perform (United States Military, 2011).

Column 3: Failure mode—As seen, failure mode is the description of how an item fails to comply with its function. It understands the events that lead to a partial or total decrease of the function of the item and its performance goals. It is recommended, whenever possible, that the FMEA process considers the history of failure of the equipment during the last years, instead of using all the possibilities of failure.

Column 4: Failure effect—A description of what happens when a failure mode occurs if no other action is taken to otherwise address the failure.

It is important to pay attention to the fact that a particular failure mode can have more than one effect, and all of them must be inserted in the table and careful not to confuse the effect with the failure mode. For example, in the FMEA of a centrifugal pump, we have the following failure mode: pump shaft rupture and the effect: The pump stops but the electrical motor continues to run, the flow of the fluid being pumped will cease, and there will be a high-level alarm in the reservoir. Cause of failure: Cracks in the pump shaft caused by poor quality of the shaft surface treatment and corrosion.

Column 5: Severity index (S)—This index is an indication of the severity of the failure effect. Consideration should be given to the effect on environment, safety, product quality, performance, loss of function and time the machine is likely to be out of service.

The criteria for assessing the severity of failure will vary according to the type of industry or process. In continuous process industries, a shutdown of certain equipment can cause the entire production line to stop, causing large production losses and even possible safety and environmental problems, as well as additional difficulties to restart the plant operation.

Moreover, for the severity index, not only its localized effect, but also the side effects in every production line should be considered.

Once the effects of the failure are evaluated, the team responsible for the analysis should use the severity table to allocate the weights for each effect (Table 4.3).

## TABLE 4.3
## Severity Index

| Rank | Criteria |
|------|----------|
| 1 | An insignificant effect not perceived by production. |
| 2 | An insignificant effect, which is perceived by production, but does not affect the system performance. |
| 3 | Negligible effect, which causes minor inconvenience to production, but does not affect the system performance. |
| 4 | Negligible effect, which causes disruption to production and slightly affects system performance. |
| 5 | Minor effect, which causes disruption to production but does not affect system performance. |
| 6 | Minor effect, which causes disruption to production and reasonably affects system performance. |
| 7 | Moderate effect, which impairs system performance leading to a serious failure or a failure that can prevent the execution of system functions. |
| 8 | Significant effect, resulting in serious failure but does not endanger the safety of people and does not result in significant cost. |
| 9 | Critical effect that causes customer dissatisfaction (production), disrupts project functions, generates significant cost of failure and poses slight safety risk (there is no risk of fatalities or serious injury) of people. |
| 10 | Very severe, risk of fatalities or serious injury or other significant cost of failure that jeopardizes the operational continuity of the organization. |

Column 6: Cause of failure—One failure mode can have more than one cause. The causes of each failure mode must be identified and listed, as these form the basis of all corrective measures. In this column only, the primary cause of the various failure modes must be filled in, through the information obtained from the maintenance records and also from the experience of the work group members.

Primary cause is the chemical or physical process, design defect, quality defect, misuse or other process that is the primary reason for the failure or that initiates the physical process that precedes the failure. Indicates why the failure mode occurred. For example: If an electric motor is turned off because the thermal protection acted, there could be several causes for the thermal relay to act. It is essential that the main cause is verified so that preventive or corrective actions are effective.

Column 7: Occurrence index or frequency of failures—Classify the failures according to their probability of occurrence with grade 1 assigned for failures with low probability of occurrence, that is, infrequent failures and, at the other extreme, grade 10 should be attributed to failures with high probability of occurrence or frequent failures.

Failure frequencies can be obtained from historical maintenance records to fill this column. However, if records are incomplete or missing, which is unfortunately the most common, staff may base their estimate of likelihood of occurrence on reports of similar components, equipment manufacturer data, or technical literature. The probability of occurrence can be classified from 1 to 10 according to Table 4.4.

### TABLE 4.4
### Occurrence Index or Frequency of Failures

| Rank | Criteria |
|---|---|
| 1 | Extremely remote chance for failure |
| 2 | Remote chance for failure |
| 3 | Minor chance of occurrence |
| 4 | Small number of occurrences |
| 5 | Occasional number of failures |
| 6 | Moderate occurrence |
| 7 | Frequent occurrence |
| 8 | High failure rate |
| 9 | Very high failure rate |
| 10 | Very high failure rate; failure is inevitable |

Column 8: Symptom of failure—A symptom is an observable phenomenon that originates and accompanies a failure or a defect. A failure symptom indicates how a failure will become apparent to the operator or maintenance personnel. These symptoms may become evident both before and after the failure actually occur.

Since the goal is to prevent failures, this step consists of deciding how likely symptoms of failure can be detected before it occurs. The information to be placed in this field will be developed in two stages: (1) consider the effects and likely symptoms of failure and (2) decide how these symptoms can be detected. For examples, refer Tables 4.5 and 4.6.

#### DETECTION OF VARIOUS EFFECTS/SYMPTOMS—EXAMPLES

Visual Detection: wear particles or other signs of damage, breaks, blockages, gaps or lack of elements, leaks, cracks, corrosion of critical elements, sparking in electrical contacts, color (oil contamination), unwanted humidity.

### TABLE 4.5
### Failure Mode and Symptom of Failure—Mechanical, Hydraulic or Pneumatic Systems

| Failure Mode | Symptoms of Failure |
|---|---|
| Malfunction or shutdown (mechanical, hydraulic or pneumatic systems) | Visible damage to machine (breakage, rupture, excessive wear); Warning signal; Excessive noise or vibration; Incorrect movement or positioning; Incorrect speed; Very high or low pressure; Very high or low temperature; Improper performance with load. |

**TABLE 4.6**

**Failure Mode and Symptom of Failure—Electrical or Electronic Systems**

| Failure Mode | Symptoms of Failure |
|---|---|
| Malfunction or shutdown (electrical or electronic systems) | Visible damage; Warning signal; Excessive noise or vibration; Intermittent operation; Incorrect voltage, current or speed; Very high or low temperature; Improper performance with load. |

Measurements: instant measurements of certain measurable parameters (physical dimensions, pressures, temperatures, noise and frequency levels, records of certain measurable parameters).

Analysis: gases, fluids, lubricants, particle deposition, frequencies, vibration, current; NDT (non-destructive testing).

Instrumentation and Control: very high or low instrument readings, unusual data in recording instruments, unusual indications in control systems, warning lamps, etc.

Sound, Noise: excessive noise, changes in noise levels or frequencies, noise from leaks (gases), warning signs, alarms.

Touch: excessive vibration, unwanted movement, high temperature.

Smell: unusual odor due to excessive temperature, friction, odor due to electrical overload, contact sparking, excessive insulation heating, etc.

Column 9: Detection index (D)—It is important to note that in the FMEA, considering the failure detection, the high values mean "not likely to be detected" and low values mean "very likely to be detected."

Another point to consider is the detection time of the failure; that is, if a failure only becomes detectable shortly before it occurs, it is a much worse situation than the failures that manifest themselves in advance, allowing the organization of tasks to minimize its consequences.

For example, an electronic component virtually fails instantly, without allowing any action to be taken to mitigate the consequence of this failure. On the other hand, some failure modes in bearings are manifested by vibration, allowing the replacement of the bearing to be scheduled in advance, minimizing the impact of the failure on the equipment and consequently on the operation.

Therefore, when selecting the classification of a certain failure mode in the detection index (D), the number of points assigned should not only reflect if the failure is detectable (operator detection or maintenance easiness), but they should also indicate if this failure provides some kind of warning with enough time for an intervention to be performed, before the failure has to minimize its consequence (Table 4.7).

As with other scales, the detection index above is only a reference and must be adjusted according to the reality of the place under analysis.

Column 10: Risk priority number (RPN)—After ranking the severity, occurrence and detection levels for each failure mode, the team will be able to calculate the

**TABLE 4.7**
**The Detection Index**

| Rank | Criteria |
| --- | --- |
| 1 | Almost certain detection of failure mode |
| 2 | Very high likelihood of detecting failure mode |
| 3 | High likelihood of detecting failure mode |
| 4 | Moderately high likelihood of detecting failure mode |
| 5 | Moderate likelihood of detecting failure mode |
| 6 | Low likelihood of detecting failure mode |
| 7 | Very low likelihood of detecting failure mode |
| 8 | Remote likelihood of detecting failure mode |
| 9 | Very remote likelihood of detecting failure mode |
| 10 | Cannot detect failure mode |

RPN, which is the result of the multiplication of the factors of Occurrence, Detection and Severity, that is to say: RPN = Occurrence × Detection × Severity.

This index is used to prioritize failure modes, that is, those with the highest indices should be treated first. It is worth noting that the RPN value has no meaning in itself. Although it is valid to say that the higher RPN values usually indicate more critical failure modes, however, caution must be exercised in the analysis.

As a general rule, any failure mode that has a Severity index of 9 or 10 must have priority. Severity Index should be given the most weight when assessing risk. Next, the Severity and Occurrence (S × O) product would be considered, since this in effect, it represents the criticality.

We present some examples of the design-out maintenance application that can be used to reduce the RPN by reducing the indexes that compose it:

- Severity: add safety devices. For example, pressure relief valves (PSV), photo-electric barriers, safety mats, dual control, and fail-proof devices. Review the materials used to manufacture the components. Limit capacity.
- Occurrence: add systems in parallel (stand-by), additional pipe support to reduce vibration, strengthening the foundation of the equipment. Check degree of protection of electric motors or install protection against leakage of liquids on electric motors.
- Detection: increase frequency of inspections, review need to add measuring instruments (pressure meters, flow meters, thermometers, etc.), check the pulley guards or any rotating component for visual inspection without disassembly.

Column 11: Recommended corrective/preventive actions—In this field, the recommended actions should be completed to prevent potential problems, or as we saw in the previous item: reduce the severity/consequences of potential problems, increase the probability of detection of potential problems (Figure 4.6).

Example of FMEA

Systems: Large Squirrel Cage Induction Motor - FMEA Number: XYZ-000　　　　Date: __/__/__

Sub-systems: _____　　　　Participants _____

| Component | Function | Failure Mode | Effect of Failure | S | Cause of Failure | O | Symptom / Detection Methods | D | RPN | Recommended Actions |
|---|---|---|---|---|---|---|---|---|---|---|
| 1. Stator Winding | Produces a sinusoidally distributed, rotating magnetic field in the stator when 3-phase ac voltage is applied to the 3-phase stator winding. | Winding-to-ground fault | Electrical trip | 8 | Thermal degradation of insulation due to high ambient temperature, restricted ventilation, under- or over-voltages, low frequency, mechanical overload, voltage imbalance, single-phasing, too frequent starting, high process fluid temp, dust or dirt accumulation | 3 | Internal visual inspection AC Hipot test Insulation power factor test Partial discharge test | 3 | 48 | Good maintenance to keep motor clean and ventilation clear and unrestricted Use surge capacitors. Good operating practices to reduce number of starts. Monitor and trend vibration. Monitor and trend ambient temp, winding temp, motor amps, rpm, process fluid temp, cooling water temp. Monitor and trend insulation condition parameters |
| 2. Shaft Assembly | Carries rotating elements of motor including rotor assembly, balancing weights, and flywheels, and transmits the torque generated by the motor to the driven load via a coupling | Misaligned | 2.1 Increased vibration 2.2 Increased wear and damage to bearings | 7 | Installation or manufacturing error Mechanical transient such as seized pump, bearing, displaced rotor bar Vibration Bowing of horizontal motor shaft | 5 | Visual inspection and alignment check Vibration monitoring Bearing temperature monitoring | 2 | 70 | Periodic visual inspection Monitor and trend vibration Monitor and trend bearing temperature |

**FIGURE 4.6**　Example of FMEA table. (Based on Wlaran & Subudhi, 1996.)

## DEVELOPMENT OF MAINTENANCE PLANS

At this stage, to complete the RCM process, the Maintenance Plans must be developed. To do so, it is necessary to select the maintenance tasks applicable to the failure modes defined in the FMEA. This activity is supported by the Decision Diagram presented in Figure 4.5.

To proceed, let's look in more detail at what is a Maintenance Plan. It is a table-format document that is used when developing the tasks necessary to properly maintain the plant equipment. The Maintenance Plan is a roadmap to help the maintenance crew to perform the necessary maintenance tasks, ensuring that the development is done consistently for all equipment.

Maintenance Plans are usually developed by the maintenance and reliability engineering team, and then they are sent to the Planning and Scheduling group to be uploaded into the CMMS, becoming a part of the planning and scheduling routine, a process that we will discuss in more detail in Chapter 7.

The following are some examples of plans: Electrical Preventive Maintenance Plan, Instrumentation Preventive Maintenance Plan, Mechanical Preventive Maintenance Plan, Inspection Plan, Predictive Maintenance Plan—Thermographic Inspection, Predictive Maintenance Plan—Vibration Analysis, Instruments Calibration Plan, Pressure Vessel Inspection Plan, Lubrication Plan, etc.

Once the plans have been loaded with the tasks that should be performed specifically, at the frequency previously determined in the CMMS, it is possible to obtain the leveling of the workload, that is the number of man-hours necessary to execute all the plans within the defined period.

The so-called "planning horizon", that is, the period covered by maintenance planning, typically is 1 year or 52 weeks.

For execution of plans, maintenance work orders are issued and they must contain the following information: Identification of equipment, detailed description of each maintenance task to be performed, the task frequency, specialty (execution team: electrical, instrumentation, mechanics, etc.), equipment condition for execution of the task (shutdown or running), type of work: Inspection, Corrective Maintenance, Predictive Maintenance or Preventive Maintenance, work procedure, estimated time to perform task, special tools, materials and support equipment (handling and lifting equipment, scaffolding, etc.) required to perform the task, Job Safety Analysis (JSA) and Work Permit (if applicable), personal protective equipment (PPE) requirements, if any.

The Maintenance Plan can be developed for each specific component of an equipment, by type of equipment or system. In general, it is best to develop plans for each class of equipment and then apply the type of maintenance identified as most suitable for all the same type or class of equipment.

When equipment of the same type is installed but in different operating environments, care must be taken to assess the impact of the operational context of each equipment, and this consideration should be reflected in the corresponding Maintenance Plan.

Aiming at the best effectiveness of Maintenance Plans, some points should be considered when designing them.

Preventive Maintenance tasks should address some mode of failure, be clear and specific and, where applicable, include specifications and tolerances.

Other points to consider to ensure that Maintenance Plans meet their objectives are as follows:

Training the performers in the specific tasks of the plans they will execute, making it clear to them what the objectives of each task are.

Implement a formal process to evaluate the adequacy and quality of the Maintenance Plan tasks. This process should establish an assessment schedule and the respective assessment report with recommendations and improvement plan.

Returning to the diagram analysis in Figure 4.7, if we look at the complete diagram with the reactive and proactive modes, we will see that there is a feedback loop linking the two halves of the diagram and it is exactly this essential part of the process that we are going to discuss now.

As we saw earlier, when we present the reactive mode, data on equipment failures should be captured in the CMMS. This data should be analyzed by the maintenance and reliability engineers who decide which are the most appropriate actions in response to each failure. To do so, the diagram shown in Figure 4.7 describes the suggested process for deciding what actions should be taken every time a new failure is identified.

This activity of maintenance and reliability engineers is essential for the improvement of the Maintenance Plans and the maintenance itself, but in practice we find that in many places these professionals, instead of dedicating themselves to these tasks, become only "crutches" for maintenance supervisors, that is, they provide technical support for supervisors who do not always have the necessary technical preparation to perform the function, or worse, become the "handyman" for the maintenance manager helping in the preparation of the most diverse types of report; even in the preparation of events like the Christmas party. The worst is that, as a rule, the maintenance manager himself is the one who deflects the focus of the maintenance engineer to "extinguish fires," helping to relieve the pressure that entire team suffers when the routine does not work as it should.

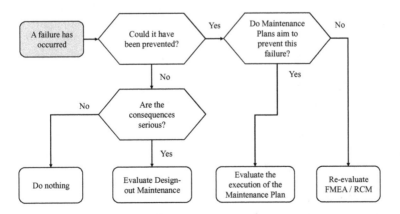

**FIGURE 4.7** PM revision decision logic diagram. (Based on Porril, n.d.)

We could compare this situation with a person who has chronic headache and simply takes painkillers. It has a temporary relief but, by not identifying and treating the cause of the pain, it returns as soon as the analgesic effect passes, that is to say, the vicious circle is formed because, if the engineer who should be dedicated to identify the failure causes and eliminate its recurrence becomes a luxury helper, the process will never be reversed and everything will remain like a "dog chasing its own tail."

With regard to the abovementioned situation, I will tell you about an experience that I had years ago when I was managing an industrial maintenance contract, where we were responsible for all the maintenance activities of an industrial plant.

At this site, our maintenance team was reasonably small and I had only one maintenance engineer responsible for all engineering activities. This professional was technically very well-prepared, extremely educated and dedicated so that our customers, especially the customer's production team, were greatly in need of the help of this engineer, even in activities that did not concern him or that were the duties of the maintenance supervisor.

Once we had an outbreak of failures related to the pneumatic automation in general; it was all kinds of problems with valves, pistons, actuators, etc. in several equipment and systems. It was a real nightmare!

The engineer, whom I mentioned, through a preliminary analysis of the failures, concluded that the problem was due to deficiencies in the system of generation and distribution of compressed air.

The problems ranged from air quality itself to the distribution system which, due to the alterations suffered by the manufacturing process over the years, would need to be completely overhauled because, among other problems, there were many piping stretches that should be deactivated as the equipment that once fed didn't exist anymore.

When he informed me of his preliminary findings, I asked him to prepare a detailed report on the compressed air generation system to be delivered to the customer since this was an investment and contractually it was the customer's responsibility for any significant overhaul in the equipment and facilities.

In order to prepare the report that I had requested, the engineer should make a detailed study of the current system and propose the modifications necessary for the system to meet the plant compressed air demand, and this would require the engineer's full dedication and, especially, time.

The great difficulty was just getting the engineer to concentrate in the study because the customers interrupted him constantly asking his help with many different things and he was embarrassed to say "no" to the customer.

The time was passing and I had to constantly justify equipment shutdowns due to pneumatic system problems.

One morning, after a major equipment shutdown due to failures in the pneumatic system, I called the engineer and asked how much time he would need to complete the study and issue the report.

He told me that if he had full time dedication in this task, in 2 weeks' time he would have the job done.

Immediately, I informed him that from that moment on, for all intents and purposes, he would be sick and would be away. I told him to go home immediately and only return with study and report ready to be discussed, before presenting to the customer.

To conclude, in less than 2 weeks the engineer "healed" (I cannot even remember what kind of illness we told the customer that he had) and presented an excellent report diagnosing the situation and proposing improvements. This report was very well accepted by the customer and most of the recommendations were implemented, significantly improving the performance of the equipment with respect to pneumatic systems in general.

## KEY PERFORMANCE INDICATORS

To complete the analysis of the diagram in the proactive mode, we will address the last missing block: the KPI.

The execution of the maintenance activities, using the techniques described in the various blocks that make up the diagram, should be evaluated and the deviations corrected in order to achieve the goals defined for maintenance. It is necessary to define what are the performance indicators that should be adopted in order to ensure that there is a direct correlation between the maintenance activity and the performance indicator that will measure it. A good test of its validity is to seek an affirmative answer to the following question: If the maintenance does "everything right," will the suggested indicator always reflect a result proportional to the change; or are there other factors, external to maintenance, that could mask the improvement?

It is worth remembering that many maintenance indicators are obtained from other basic technical and economic indicators. Therefore, it is very important to make sure that the organization captures the appropriate data for the calculation of these indicators in order to meet the level required for the performance analysis in question.

Looking at the maintenance process described in the diagram as a whole, its main objective should be to provide the performance required by the company to meet its corporate objectives and each element within the maintenance process is itself a subprocess.

One of the definitions of the maintenance mission is to ensure the reliability and availability of the assets in order to meet a production schedule with safety, environmental preservation and adequate costs. Therefore, we can conclude that if maintenance is fulfilling its mission, it will be fulfilling its role in the company's corporate objectives.

In this way, the indicators to be defined should reflect: Reliability, Availability, Maintenance Costs, and Safety.

Reliability: measured by the Failure Rate ($\lambda$) and MTBF (Mean Time between Failures) of equipment or systems.

Failure rate ($\lambda$): number of failures observed in a given time interval.

They are usually referred to as constant failure rate because they do not vary over time. The inverse of the failure rate provides the MTBF or MTTF (Mean Time to Failure).

Availability: Remembering the definition presented in Chapter 2, we can say that the availability, for a certain period of time, measures the percentage of time the equipment was available to operate in relation to the total time.

It should be noted that some sources consider this definition as operational availability because it is a measure of the average availability over a period of time and that includes all sources of equipment inactivity, such as administrative downtime and logistic inactivity time.

This definition includes both downtime for corrective and preventive maintenance. Because the operational availability is the real level of availability in the day-to-day operation of the plant, we understand that it is the indicator to be considered in order to better evaluate the maintenance performance.

We will detail the factors that compose this indicator through Figure 4.8.

Analyzing the graph below, the period from point 0 (the instant when the failure occurs) to point 1, which corresponds to the beginning of the repair, is called MWT or Mean Waiting Time.

This period includes the following times: time to find the maintenance professional who will perform the repair, travel time and the Work Permit and release of equipment for repair (if applicable).

The high MWT can reveal management problems such as organization of the maintenance team, communication system, bureaucratic process and inadequate logistics.

The period ranging from 1 to 2 is the repair time (MTTR), and this time could be further subdivided as follows: the time to identify the cause of the failure and propose the correction; the time to provide spare parts, tools and ancillary equipment to do the repair (logistic time); and the time for repair itself and test execution.

High MTTR may represent problems with staff qualification, maintenance issues (see Chapter 2), logistics problems, problems with spare parts and maintenance materials management, etc.

MTBF, as we have seen, is directly connected to the reliability of the equipment and should be the first to be evaluated because, if there are no failures, none of the other factors will exist.

In practice, the major problem is the identification of these times because, in the absolute majority of the places that I have known around the world, this type of data is not recorded in the maintenance Work Orders, nevertheless, this set of indicators are essential for the maintenance performance analysis and the factors that affect it.

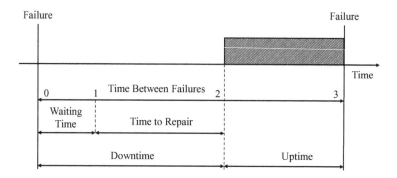

**FIGURE 4.8** Uptime and downtime details.

Total Maintenance Cost: simply stated, maintenance costs include direct labor with overhead and benefits, materials, contracted labor and services, and others. The detail of the main items of each one of them is as follows:

- Cost of labor including wages, salaries and overtime for managerial, supervision, support staff and direct staff; also payroll added costs for the above-mentioned persons (taxes, insurance, legislative contributions). In general, overtime should not exceed 5% of the maintenance labor cost, and the total cost of labor typically costs around 45% of the total cost of maintenance.
- Cost of materials including the spare parts, generally materials that are used in the manufacture of parts or installations. Typically, this cost represents something around 35% of the cost of materials.
- Cost of contracted labor and services: examples include rent of scaffolding, load handling, machining, predictive maintenance services, analysis of insulating oil and lubricant, thermal insulation services, electric motor repair, instrument calibration, painting and specialized services performed exclusively by the equipment manufacturer. This item accounts for something around 20% of the total cost of maintenance.
- Others: we can exemplify some expenses, usually of lesser impact, such as replacement of tools (which are not considered fixed assets according to tax legislation), consumables (sandpapers, adhesives in general, duct tape, degreaser, brushes, electrodes, paints and varnishes, industrial gases, solvents, office materials, PPE—personnel protective equipment, etc.)

In addition to the above indicator that shows the absolute cost, there are other indicators that present the maintenance cost as a percentage of the Asset Replacement Value (ARV).

The ARV can be defined as the amount of capital that would be required to replace the asset. This is not the book value nor the current cost accounting value nor the cost to build a state-of-the-art replacement. ARV is an estimate of the current costs to replace in kind what now exists (ARV can be equivalent to the insurance value).

The advantage of using maintenance cost related to ARV is to be able to compare its results with those obtained by leading companies in order to firstly locate and then perform an analysis of benchmarking opportunities.

As we said earlier, maintenance should be aligned with the company's strategic objectives and, certainly, profitability is one of the primary factors for its survival. In this way, maintenance costs play an important role in the profitability of the business, since profit is defined as the difference between the sales price and the cost to produce and sell.

It should be noted that maintenance costs are impacted by both the maintenance strategy and the efficiency with which the tasks are performed. For example, in implementing the predictive maintenance using vibration analysis, the potential failure of a bearing can be identified and, thus, the intervention can be planned and scheduled, and this will entail costs much lower than the costs resulting from a catastrophic failure of the same bearing, if the maintenance type was "run to failure."

Catastrophic failure of a motor bearing can even cause irreversible damage to the rotor/stator assembly of the motor and consequently total loss of the motor.

Maintenance costs are also impacted by low maintenance productivity, which can be divided into three factors, namely:

- Labor Utilization Factor (lost time): including planning and preparation, waiting time (instruction, tools, materials, release, supports), travel time and idle time.
- Performance Factor (the speed to execute the task) which will depend on the qualification of people (knowledge, skills and experience) and the tools and working methods.
- Quality Factor (unnecessary tasks) which is the quality of the task execution (avoiding rework), the measures to prevent recurrences of equipment failures and the equipment design improvements to eliminate tasks.

Many companies have their maintenance productivity (also known as "wrench time") with values between 25% and 30%. From our own experience, we have seen a company increasing the maintenance productivity from 26% to 50% through a well-implemented project, involving productivity measurement by Work Sampling and subsequent implementation of identified actions for improvement.

In addition to this "visible" side of the maintenance cost, we have talked about so far, there is another cost for which maintenance is responsible, which is not always visible to all, which is the losses resulting from the total unreliability of the equipment or its performance reduction. In other words, the losses associated with the production shutdown, when the equipment is not available because of maintenance interventions, either corrective or preventive.

These costs are often compared to an iceberg (see Figure 4.9) because direct costs are the visible part of the iceberg, the one that is above the waterline while the

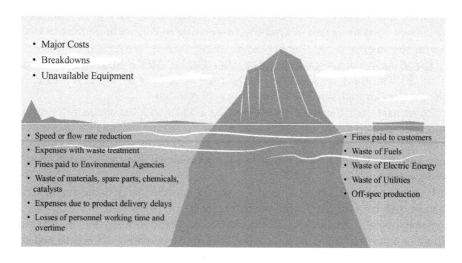

**FIGURE 4.9**   OEE iceberg.

indirect costs, which make up the bulk of the total cost, would be the submerged part of the iceberg, that is, the part that is not visible.

Not many companies are able to answer accurately how much it costs an hour to stop a particular equipment or even an entire production line, and this information is extremely important as it can affect the strategy of maintenance and assist in demonstrating that blindly reducing the cost of maintenance, as many companies do in times of crisis, can, at the limit, not only affect the operational continuity of the company but also go against its strategic objectives.

There is a demonstration using the so-called "DuPont Model" technique to locate the areas responsible for the company's financial performance to show the impact of Overall Equipment Effectiveness (OEE) on the company's financial results.

I think it's worth looking at this demonstration because it can help us understand how the relationship between direct maintenance costs and financial losses by reducing availability works.

First of all, for a better understanding of the demonstration, we will review the OEE indicator. This indicator was developed in Japan to, in a simplified way, evaluate the effective use of equipment, that is, how many good products were produced, compared to the amount of total goods that could have been produced.

The OEE is composed of three factors: availability, performance and quality.

OEE = Availability × Performance × Quality—in Figure 4.10, we see what influences each of the factors:

To illustrate the use of OEE, let us imagine that a company has the capacity to produce 1,000 pieces per month and has the following losses:

- 100 pieces were not produced due to equipment failure.

$$\text{Availability} = \left(1 - \left(100/1,000\right)\right) = 90\%$$

- 40 pieces were not produced because of reduction of the line speed.

$$\text{Performance} = \left(1 - \left(40/900\right)\right) = 95.6\%$$

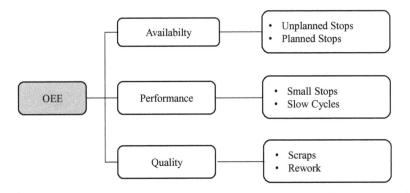

**FIGURE 4.10**   OEE definition.

- 60 pieces were scrapped due to quality issues.

$$\text{Quality} = \left(1 - \left(60/860\right)\right) = 93\%$$

- $\text{OEE} = \left(1 - \left((100 + 40 + 60)/1{,}000\right)\right) = 80\%$ or $90\% \times 95.6\% \times 93\%$

From the review of the OEE concept, we will analyze the example shown in Figure 4.11.

One company operates with an OEE=60%, and we will focus our analysis on the effect of changing the plant Availability, one of the components of OEE, and this one measures 75%.

From this OEE, net production is 262,800 good parts per year, and these parts are sold for $ 600.00 per piece with a profit margin of $ 250.00 per piece that results in annual profit of $ 11.2 million that produces a return on the capital employed (ROCE) of 6.75%. Note that in this first situation, the annual direct cost of maintenance is approximately US $ 9.5 million.

Now, we are going to observe a new hypothetical situation where it was invested in the maintenance being possible to obtain considerable improvements in the

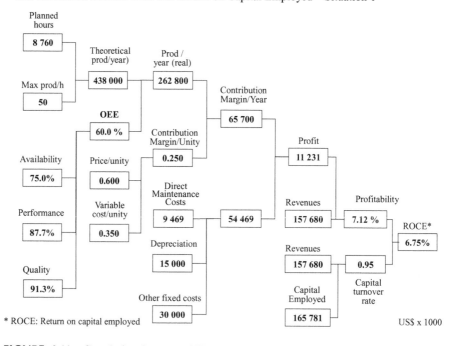

The correlation between OEE and Return on Capital Employed – Situation 1

**FIGURE 4.11** Correlation between OEE and return on capital employed—Situation 1. (Based on Leroux, n.d.)

The correlation between OEE and Return on Capital Employed – Situation 2

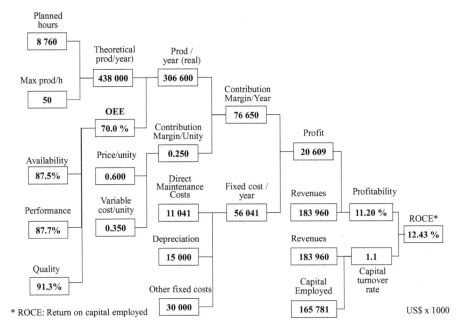

**FIGURE 4.12** Correlation between OEE and return on capital employed—Situation 2. (Based on Leroux, n.d.)

availability and it went from 75% to 87.5% and the other factors remained the same of the previous situation[1] and the OEE is now in 70%.

With this considerable increase in OEE, net production increases to 306,600 pieces, and assuming that the market absorbs all this increase of production, that is, that all production is sold, then the annual profit becomes US $ 20.6 million and, in this situation the ROCE becomes 12.43% and the profitability that was 7.12% came to be 11.2%.

It is worth noting that in this new situation, the annual maintenance cost even increased to $ 11 million (see Figure 4.12 with full analysis).

Let us now imagine a third hypothetical situation where we want to obtain the same profit and ROCE from the previous situation; however, instead of improving the Availability we will try to reach the same results by reducing the direct maintenance cost. Figure 4.13 illustrates this third situation and the resulting values of this simulation.

The immediate conclusion of this exercise is obvious, that is, we would practically need to reduce the direct maintenance cost to zero to achieve the same profitability obtained with a 12.5% increase in availability (in the case of this example).

---

[1] We set the values of Performance and Quality, although this situation hardly occurs in practice, in order to isolate the influence of Availability on the profitability that is the object of our study.

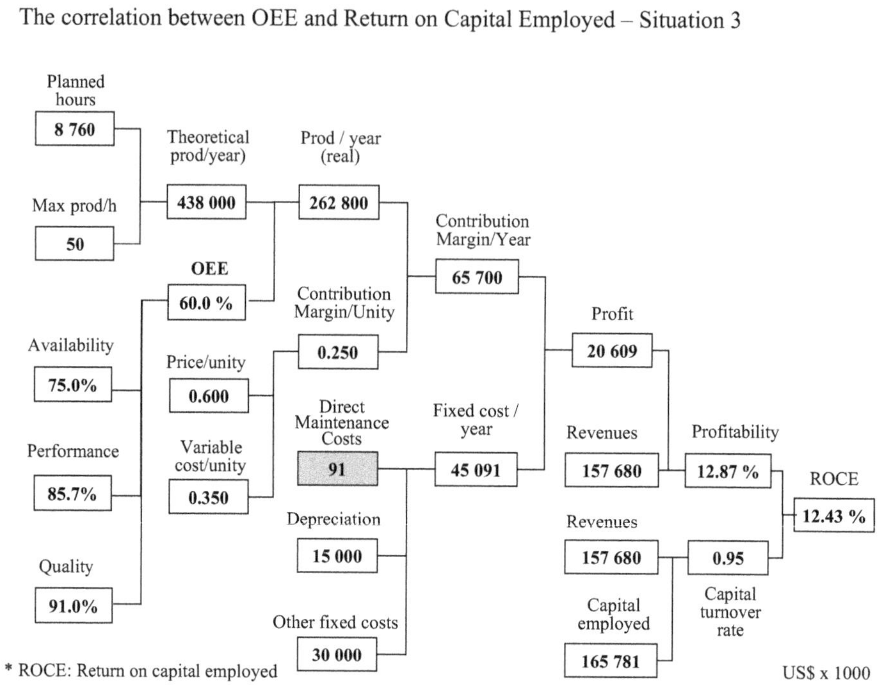

The correlation between OEE and Return on Capital Employed – Situation 3

**FIGURE 4.13** Correlation between OEE and return on capital employed—Situation 3. (Based on Leroux, n.d.)

Our objective with this example was to emphasize the importance of knowing the impact of availability on the company's profitability and carefully evaluating any and all cost-cutting action to avoid unwanted or even disastrous consequences.

## BENCHMARKING

To conclude this topic, it is important to mention the Benchmarking process, that is the process of comparing results obtained with results of other companies that are leaders in their segments.

Table 4.8 shows indicators that are common in industrial maintenance benchmarking. The presented values were established by the board of directors of the North American Maintenance Excellence Award as representative of superior maintenance operations. The presented indicators are not intended as targets to be met or guidelines to be followed in maintenance evaluation.

## TABLE 4.8
## Industrial Maintenance Benchmarking

| Benchmark | Process Industries | Discrete Manufacturing |
|---|---|---|
| Maintenance technicians per supervisor | 15 | 16 |
| Maintenance technicians per maintenance planner | 28 | 25 |
| Plant replacement value per maintenance technicians (in USD millions) | $4.9 | $4.5 |
| Training days per maintenance technician (annual) | 9.3 | 9 |
| Training days per maintenance supervisor (annual) | 7.4 | 5 |
| Total maintenance cost as percent of estimated plant replacement value | 2.2% | 2.5% |
| Maintenance labor cost as percent of total maintenance cost | 24.5% | 52% |
| Contracting cost as percent of total maintenance cost | 20.3% | 15% |
| Unplanned (emergency) man-hours as percent of total maintenance man-hours worked | 7% | 18% |
| Preventive/predictive man-hours as percent of total maintenance man-hours worked | 27% | 39% |
| Planned corrective maintenance schedule compliance | 87% | 87% |
| Preventive maintenance schedule compliance | 90% | 90% |
| Work order coverage as percent of man-hours worked | 100% | 98% |
| Storeroom annual inventory turnover | 1.52 | 1.53 |

Based on Dunn, 2001.

# 5 Maintenance Basics
## *The Forgotten Cornerstone of a Solid Maintenance Foundation*

Once during a visit to South Africa, where I was participating in the start-up of a maintenance management contract, I was asked by local colleagues to attend a business meeting with one of their potential clients. They told me that they informed the client about my visit to the country and the client expressed an interest in meeting me to get insight into our maintenance management process because they were discussing a possible outsourcing of a maintenance contract and, at that time, I was globally responsible for the reliability and maintenance process of the company.

At the meeting, as soon as the formal presentations were made, the plant manager asked me the following:

"As a specialist, what would you do to improve the performance of this plant in terms of reliability?"

I had not even stepped into that plant and also had no idea about the conditions or performance indicators; however, avoiding evasion or even rudeness, I replied:

"Obviously, it would be frivolous of me to make any specific comments about your plant since I did not visit it and received no information about its performance. As we know, reliability performance is directly linked to the occurrence of equipment failures. We can state, according to the results of studies, that most failures are due to the negligence of basic equipment conditions such as: cleaning, lubrication and fastening. It should also be noted that operating and environmental conditions must obey the design specifications of the equipment, so if this plant has poor performance and a high number of failures, I would start by examining these conditions I have just mentioned. And that's all I can say without a detailed assessment."

At the moment of the question, I noticed a certain tension and anxiety in the face of my colleagues because, depending on my answer, we could disappoint the client, damaging the commercial process that was in progress and the visit that should be to promote the company could have the opposite effect.

Upon hearing my response, the client smiled and said he fully agreed with my opinion, expressing his satisfaction at choosing to start with the basics before talking about new technologies or advanced science.

Strictly speaking, this chapter should have been the first, but because I chose to develop the "case" I related, I left it to put it in this sequence without prejudice to its order of implementation or importance.

Going back to my answer to the customer in South Africa, it is based on the concept of Total Productive Maintenance (TPM) called "Zero Breakdown" which, according to the terminology we are using would be "Zero Failure."

Based on Suzuki, 1994, one of the objectives of TPM is to eliminate the equipment failures ("Zero Breakdown") and, to reach this goal, some measures need to be taken, starting by restoring the basic conditions of the equipment through cleaning, lubrication and fastening. This measure aims to eliminate the equipment accelerated deterioration.

It should also be ensured that the equipment is being used in accordance to its specified operating conditions.

The process should be restored to its ideal conditions, eliminating the environmental conditions that cause the accelerated equipment deterioration.

The next step is to extend the equipment life by correcting any possible design deficiencies.

And, finally, unexpected failures can be eliminated through improvements in the equipment operation and maintenance.

Our focus will be the first measure to be taken, and we will start by examining each of the basic conditions as follows.

## CLEANING

Someone may ask: "It's okay that the equipment cleaning affects the appearance of the premises and makes a good impression on visitors, can avoid contamination of the products, facilitates inspections and may even improve employee morale, but how does cleaning contribute to eliminate the equipment failures?"

In terms of the manufacturing environment cleaning, I can say that I had the opportunity to try heaven and hell. The heaven was when I worked in the semiconductor industry that manufactured electronic components, an environment that is called a "clean room," but in this company I was a process and project engineering manager and was not directly involved with the equipment maintenance. However, what I consider a true stage in hell, I have been working for many years in the maintenance of tire and rubber plants, where one of the raw materials is the Carbon Black. It is light coal, finely pulverized, obtained by the incomplete combustion of organic compounds and is used to improve the mechanical properties of rubber, especially its resistance to wear. Carbon Black, despite all the efforts and the dust-collecting systems installed, at the time, contaminated the whole installation around it, and the results were seen in the useful life of equipment and components and by the unfortunates who worked in that miserable place.

Returning to the question of the reasons that lead to cleanliness to reduce the equipment failures, let's show some practical examples. First, improper cleaning can cause equipment failure since dirt and foreign material can penetrate bearings, rotating and sliding parts, hydraulic and pneumatic components, electrical panels, sensors, etc.

Another problem caused by lack of cleaning equipment, perhaps not as obvious as the one mentioned above, is the deficiency in heat dissipation. Let's look at the case of an electric motor. Assume that this motor is installed in a heavily contaminated area and that it has dirt in its entire housing. Consequently, in addition to the motor surface being thermally insulated, the air inlet located on the rear cover is also partially blocked by the dirt accumulation. In this way, the motor will be less able to dissipate heat and will operate at a higher temperature than it should. According to the manufacturers of electric motors, the service life of the motor is inversely proportional to the operating temperature, in other words, as the operating temperature increases, the motor life decreases. WEG, a traditional manufacturer of electric motors, in one of its manuals, concluded that for rugged, simple machines such as induction motors, their life span depends basically on the winding insulation and its service life. The winding insulation service life depends on many environmental variables, such as humidity, corrosion, dust and others, being temperature the most important by far. For example, winding life span can be decreased in 50% with a shift of 8–10° above the thermal class limit (Grupo WEG, 2017). Obviously, one motor operating in these conditions will not fail immediately but will have a significantly reduced service life, as per the manufacturer's own statement.

The second case to be analyzed are the hydraulic systems. Hydraulic systems have reservoirs, and these are designed with sufficient surface area to dissipate the heat generated during their operation. Hydraulic equipment manufacturers recommend that tanks be installed in a covered, airy, clean, dry location in an environment with the minimum of dust suspended in the air, and away from sources that radiate heat. This allows the hydraulic unit to have good heat exchange with the environment and low probability of accumulation of dirt on its surface. If the installation does not comply with these recommendations, dirt or contamination build up on the reservoir surface, making thermal exchange difficult and, in this way, the hydraulic fluid retains heat, raising the working temperature of the entire system causing damage to the system, such as oxidation of hydraulic fluid, which will shorten the life of the fluid, vulcanization of valve seals, pumps, cylinders and others.

## LUBRICATION

I believe that if we ask most people, who are not experts in the subject, what means lubrication, we should get in response that lubrication is something like: "the act of applying an oily or greasy substance with the aim of reducing the friction between parts." This is a valid definition, though simplistic, but it is not a complete definition since lubrication goes far beyond this.

In fact, although reduction of friction is the main objective of lubrication, there are other benefits of this process, not so evident, for example, lubricating films can help prevent corrosion by protecting the surface against water and other corrosive substances. In addition, lubrication plays an important role in controlling contamination of systems because the lubricant acts as a means of transporting contaminants, leading them to filters where they are removed. The lubricant also helps in temperature control by absorbing heat from the surfaces and transferring to a lower temperature point where it can be dissipated.

The next question would be: how can we define a correct lubrication process? And the answer is the basic principle of lubrication, that is, use the correct lubricant in the correct amount, in the correct place, at the appropriate frequency and using the proper method.

On this principle, we can clarify a little more the care that must be taken to have a good lubrication process:

- the lubricant specification is correct;
- the quality of the lubricant is controlled;
- the equipment manufacturer's recommendations are followed;
- there are no errors of application;
- the handling, storage and storage system is correct;
- the lubricator (professional who performs the lubrication) is qualified;
- the implementation procedures are elaborated, implemented and followed;
- there is a regular inspection in the reservoirs;
- the centralized systems are correctly designed, maintained and regulated;
- the scheduling of the lubrication services, with the recommended lubrication frequencies, being duly attended;
- the lubrication team is correctly dimensioned;

In theory, this seems to be very simple and easy to achieve, but in practice, experience shows us that it is not the case, quite the contrary. Most of the time, the importance of lubrication is underestimated, and the attention it deserves is not considered because of the damage that the process can cause to a company.

## THE CASE OF THE ELECTRICAL MOTOR LUBRICATION

At this point, I will open a parenthesis to tell a case that exemplifies the importance of lubrication. In one plant where I worked, one of the most critical equipment was powered by a 1000 HP medium-voltage induction motor. Once, the regular predictive maintenance (vibration analysis) recommended the replacement of one of its bearings, in the shortest possible time to avoid catastrophic failure of the motor.

As there was no spare motor, the replacement of the bearing implied to a plant shutdown because the stop of this equipment caused the total stop of production. In this way, due to the importance of the equipment and not to take any risk, we decided to hire the motor manufacturer's own field maintenance services to repair the motor on site during a forced plant shutdown scheduled for an extended weekend (Friday, Saturday, Sunday).

Unfortunately, it is very common to find a plant layout that didn't take in consideration that some equipment will need to be removed from the place where they are installed to be repaired. Because of this inconsequential practice, the motor we are talking about was installed in a place very difficult to reach, and this entailed serious difficulties in performing the repair and, consequently, consumed much more time than would be normal if the motor were installed in an easily accessible location.

As we had other services in progress, I did not follow closely the execution of the motor bearing replacement; I just asked to be called when the service was completed and the motor ready to be tested.

The motor test without load was scheduled to happen late Saturday afternoon and so it happened, with the motor manufacturer's maintenance technicians monitoring some motor parameters, such as electric current and bearing temperatures.

The facial expression of the technician, who checked the motor bearing temperature, showed that what he was observing on the instrument was not correct and this was confirmed when he asked nervously to turn the motor off immediately.

Then, he informed us that something was wrong because the bearing temperature had risen suddenly and was above normal.

Still according to him, they would need to check the bearing assembly completely to determine the possible causes of the mentioned abnormality and, as they had scheduled to finish the service on Saturday, they would need to have their company's authorization to continue working on Sunday.

The technicians contacted their superior and it was decided that they should continue the service on Sunday morning because they suspected some irregularity with the supply of grease in the bearing and would prefer not to remain in the plant on Saturday night. From our side, we warned them that production was scheduled to restart on Sunday at 11:00 PM and the motor should be able to run then as this was our commitment to the plant manager.

On Sunday the motor was opened, checks were made and nothing wrong was found and the technician, who led the team, informed us that we should start the motor without load again and observe the temperature. Now, we argued this recommendation because if nothing had been found, and consequently nothing had been done what would be the chance of having different results from those that had been obtained the first time?

The impasse was created, but since we had no alternative and we needed the motor and they were the "experts," we decided to follow their recommendations.

As expected, after the motor rework, the results were repeated, which left the technician visibly nervous and totally disoriented. He asked us some time to think about what could be done and tried to call some people of his company to ask for guidance and help, but since it was Sunday and the cell phones still did not exist at that time, he could not get any contact.

Then the technician suggested to run the motor to verify at what value the bearing temperature would stabilize and it was done. As the temperature, according to the technician, reached the maximum allowed value and stabilized, he suggested that the equipment operated in these conditions from Sunday to Monday morning to avoid affecting the production schedule. On Monday morning, he would seek advice on what steps should be taken to correct the problem.

In this way, the equipment started the operation at 11:00 PM on Sunday and we, along with the technicians, stood at the motor checking its bearing temperature as a mother watching over the child with a fever in the bed and, to everyone's despair, as soon as the equipment was loaded with raw material, the bearing temperature began to rise rapidly, exceeding the maximum recommended value.

The technician asked us to reduce the load on the equipment to check the temperature behavior and, after a few attempts, concluded that the equipment could only operate at 50% of its rated capacity to avoid further motor damage.

The equipment operated with this limitation, affecting the production schedule until Monday morning. After contacting the motor manufacturer experts, the field service technician verified that the grease with which he had lubricated the bearings of the motor was not the specified grease, according to the motor maintenance manual.

This example makes us reflect the following: if a technician, motor maintenance specialist of a reputed motor manufacturer, makes such a mistake with lubricants what can happen in the daily maintenance activities in companies that assign the equipment lubrication to totally unprepared people?

How many different types of lubricants from different manufacturers exist in a typical installation? Usually there are many. The lack of organization, defined and implemented lubrication process and well-trained professionals certainly will seriously compromise the life of the equipment as well as the total maintenance costs.

Another commonly overlooked aspect of lubrication is the use of grease guns. A grease gun in the hands of an untrained lubrication technician will certainly compromise the reliability of the equipment due to the introduction of early equipment failures.

If you ask to a lubrication technician: "How do you know how if you have enough grease in your bearings?" And he or she answers:

"When the lubricant runs out the seals."

Unfortunately, this is a common and wrong answer!

Most rolling bearings are made to only have about one-third of their volume filled with grease, leaving enough room for heat exchange. The excess of grease volume in a bearing housing can lead to high operating temperatures.

Seal damage is another negative side effect of over-greasing. Grease guns can produce up to 15,000 psi, and when someone over-grease a bearing housing, the lip seals can rupture, allowing contaminants such as water and dirt to gain access into the bearing housing.

Contamination of lubricants is another important cause of equipment failure, with the most common contaminant being water. If the lubricant contains a water content of only 0.002%, the life of the bearing can be reduced by up to 80%. In addition to water, there is also contamination of the lubricant due to solid particles that are responsible for wear and failure of the components.

Some care must be taken to avoid the lubricant contamination such as:

- Ensure that lubricant is properly stored, preventing the oil drums remaining outdoors, especially in the horizontal position, causing water to accumulate on the surface of the drum, leading to corrosion, metal wear and contamination of the oil.
- Even when taking the best care possible to store lubricants, they are subject to contamination ingression. It is recommended that the lubricant be filtered prior to entering the equipment.
- Observe the correct operating temperature of the equipment.

Reinforcing what it was said before about the importance of lubrication, according to Bannister, 2014, Dr. Ernest Rabinowicz from Massachusetts Institute of Technology (MIT), conducted studies that showed that effective lubrication practices can prevent 70% of bearing failures; mechanical wear represents 50% and corrosion 20%.

In summary, there are sufficient reasons justifying investing the time and resources to have an effective lubrication program.

## FASTENING

Fastening is another basic condition of the equipment that contributes significantly to the occurrence of failures and, like the others previously mentioned, is commonly ignored in most companies.

Mechanical equipment generally has several fasteners such as nuts, bolts, stud bolts, and anchor bolts, which are essential parts of its structure. These devices will only operate correctly if these fastening elements are properly adjusted.

Just one loose bolt can start a chain reaction of wear and vibration. If the machine vibrates slightly, other screws begin to loosen; vibration causes more vibration and the equipment begins to shake and make noise, the slight cracks become deep cracks and some parts end up damaged, culminating, often, in a great failure.

Who has had the experience of being with some equipment and having the clear sensation of listening to a drumming show?

Usually, failures and other problems are the result of a combination of conditions acting together. For example, a photoelectric cell will probably continue to function satisfactorily, even if with some vibration or with its glass cover slightly dirty. However, if the screws are loose, the increase of vibration will loosen the screws and further increase the vibration. Under these circumstances, any slight misalignment or contamination of the light receiver may then cause a system failure.

It is worth mentioning that the loosening of the screws themselves is not a problem but this may trigger a series of events that will cause the failure itself.

Failure analysis performed at an industrial plant revealed that the undue tightening of the bolts contributed, in each period, in one form or another, directly or indirectly, to approximately 50% of the equipment failures. Therefore, it is advisable to re-evaluate the importance of tightening bolts and nuts and, consequently, to take control of the torque applied to them.

Lack of torque control can cause various types of failure. For example, when the torque is too high, the following faults can occur: the screw has its threads screwed down, the bolt is broken, the bolts are screwed together, the seals are crimped, etc. On the other hand, when the torque is insufficient, the bolt may come loose due to vibrations of the equipment itself, compromise the seal of a joint causing fluid leakage, compromise the operation of the equipment due to the misalignment of components, etc.

Important to notice that it is not recommended to start any preventive or predictive maintenance program, before addressing the equipment basic conditions issues (cleaning, lubrication and fastening). These conditions will most likely cause failures before the next service is due (Suzuki, 1994). It is just a waste of time and money!

# 6 Maintenance Professionals
## *The Stigmatized Heroes*

## THE HUMAN ERROR IN MAINTENANCE

Imagine the following situation: In a chemical plant with several factories, all industrial maintenance was carried out by a specialized company under a global maintenance contract. On a certain day there would be a meeting between the directors of the companies; the manufacturer and the service provider.

The meeting was an important milestone in the history of the contract because, after several years, it was the first time that nobody less than the country manager of the service provider, a large multinational company with multiple business segments, visited the customer. The maintenance team members were aware that they were rendering a good service and that the customer was generally satisfied. However, they were still eager to know what the customer would say to their company's chief executive in South America and what impression would he take back home because, as the saying goes: "one never gets a second chance to make a good first impression."

The meeting was good with an atmosphere of cordiality between the parties. The customer, although restrained, expressed satisfaction for the services he received and pointed out that one of the great differentials was that the maintenance team was very experienced and committed and that therefore, there were no major problems in manufacturing.

At the end of the meeting, when the participants were already in front of the building for the parting, there was a certain alarm when they saw black smoke coming out of the huge chimney of the boiler building that generated steam for the whole site.

The chairman of the service provider called the contract manager aside and quietly asked, "What's happening?" And got a laconic answer: "probably, the boiler has shut off, but the reason is not yet known." The president dismissed them, requesting: "Please keep me informed and let me know if we have responsibility on this event" and left.

What happened was that the boiler shut off affecting the production of several factories of the plant: a real disaster!

And worst of all, maintenance was solely responsible for the event. The electrical maintenance crew had treated the oil from an energized transformer which fed the Boiler Auxiliary Services Motor Control Center and one of the electricians participating in the intervention, which had been commissioned to reconnect the cables of a protection relay, made a wrong decision to do one more test with a multimeter and ended up unduly activating a protection circuit, turning off the transformer and everything else it powered, causing this embarrassing incident.

Disregarding the hypothesis of sabotage, we can classify the cause of this incident as a human error.

In Chapter 3, when we discussed the causes of the so-called "infant mortality," that is, that initial period of the conditional probability of failure curve, which describes the reliability of a component as function of time, we know that one of the causes of the high initial failure rate is the introduction of equipment failures by the maintenance team itself, especially when performing unnecessary tasks. Remember the phrase: "Every time the maintenance staff do preventive maintenance on this equipment, it doesn't operate anymore...."

This phrase, which in the ears of maintenance professionals sounds like pure discrimination or prejudice, becomes more meaningful from the long list of catastrophic failures, in the most diverse types of segments, in which the inadequate execution of maintenance tasks was indicated as being directly or indirectly responsible for the occurrence.

I believe that there is no doubt that this problem deserves attention and must be addressed by those who are responsible for the maintenance management with the same dedication and depth that the reliability of the equipment is treated.

Before proceeding, in order to avoid any confusion, I would like to present some important concepts and definitions on the subject.

We will start by briefly introducing the concept of the human reliability, which is a logical consequence of the study of the reliability of the equipment, recognizing that the human being fails and that these failures can be classified, quantified and analyzed mathematically by means of statistics.

Human reliability can be affected by many factors such as age, state of mind, physical health, attitude, emotions, propensity for certain common mistakes, errors and cognitive biases.

The human failures can be classified into errors and violations that occur during the execution of maintenance tasks (please refer to Figure 6.1).

## HUMAN ERRORS

The human errors can be classified as slips, lapses and mistakes, and they constitute the main part of human failures because they depend, fundamentally, on the ability to perform the task, the stress, the environmental and social conditions, motivation and other factors.

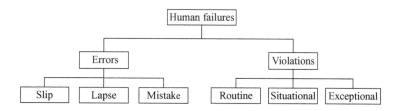

**FIGURE 6.1**   Human failures. (Based on Health and Safety Executive, n.d.a.)

Slips are failures in carrying out the actions of a task. They can be described as "actions-not-as-planned." Some examples: move a switch up rather than down (wrong action on right object), take reading from wrong instrument (right action on wrong object), picking up the wrong component from a mixed box.

Lapses: This is an error due to distraction of forgetfulness, causing us to forget to carry out an action, to lose our place in a task or even to forget what we had intended to do. For example: Short-term memory lapse; omit to perform a required action: forget to tighten the flange bolts, forget to turn off the main switch before opening the panel, miss crucial step, or lose place, in a safety-critical procedure.

Mistakes: They can be defined as a correct implementation of the wrong action. They can be divided in two different categories:

Rule-Based Mistakes: If behavior is based on remembered rules and procedures, mistake occurs due to misapplication of a good rule or application of a bad rule. For example: not ensuring that the equipment is switched off before performing the intervention, ignoring alarm in real emergency, and following history of spurious alarms.

Knowledge-Based Mistakes: Individual has no rules or routines available to handle an unusual situation, resorts to first principles and experience to solve problem, relies on out-of-date electrical schematic for fault diagnosis or diagnose equipment failure, and takes inappropriate corrective action (due to lack of experience or insufficient/incorrect information, etc.)

## VIOLATIONS

The violations (or noncompliance) are deliberate deviations from rules, procedures, regulations, etc. They can be divided into the following categories:

Routine: Violations become the "standard"; a systematic behavior opposite to the rule or procedure. For example: employees don't wear the required personal protective equipment (PPE) in the production area, Work Permit routinely issued without physical, on-plant checks.

Situational: These violations of the procedure occur when there is a restriction due to the environment, in a physical or organizational way, for example, time pressure, workload, unsuitable tools and equipment, and weather. For example: the worker worked overtime beyond the legal limit to repair broken critical equipment.

Exceptional: These are violations of the procedure for a particular situation, like an emergency such as to postpone a scheduled safety drill to prevent loss of production and to accomplish production plan.

The question that come to our minds is: what can be done about the error management since the human condition is not changeable, but the work conditions are, as stated by Reason and Hobbs, 2003.

First of all, it is important to mention that some conditions foster the occurrence of errors, among which we highlight: work environment (extremes of heat, humidity, noise, vibration, poor lighting, restricted workspace); high workload, tasks demanding high levels of alertness, jobs which are very monotonous and repetitive, situations with many distractions and interruptions, insufficient staffing levels, demanding work schedules, individual stressors (lack of training and experience, fatigue, reduced alertness, family problems, illness, abuse of alcohol and drugs) and equipment stressors (poorly designed displays and controls, inaccurate and confusing instructions and procedures).

The mentioned conditions that influence on peoples' behavior and performance need to be considered to address the human errors including:

- Design equipment and system considering possible slips and lapses occurring or to increase the chance of detecting and correcting them;
- Make sure that required training programs are effective;
- Avoid the need for tasks which involve very complex decisions, diagnoses or calculations, for example, by writing procedures for rare events requiring decisions and actions;
- Ensure proper supervision especially for inexperienced team members, or for tasks where there is a need for independent checking;
- Make sure that procedures and instructions are clear, concise, available, up-to-date and accepted by users;
- Consider the possibility of human error when undertaking risk assessments;
- Monitor that measures taken to reduce error are effective.

Based on Health and Safety Executive, 1999

Let us present a simple example to illustrate a situation that can be considered a relatively routine maintenance activity: imagine a long bolt with eight washers inserted in it (see Figure 6.2).

Imagine that this assembly is part of a certain equipment and, during a maintenance intervention, this part of the equipment needs to be disassembled and all the washers must be removed from the bolt.

It is totally unnecessary to mention that there is only one way to remove the washers from the bolt because all the steps of this task are intuitive and all the knowledge necessary to perform the task is available in the task itself, thus there is no need for the worker to remember or follow any procedure.

However, once the task is completed, imagine that the equipment should now be reassembled, that is, to replace the washers on the bolt, but there is a requirement

**FIGURE 6.2**   Bolt and washers. (Based on Health and Safety Executive, n.d.b.)

that they need to return exactly in the same position as they were before being removed.

If to remove there was only one possibility, now the situation is quite different because there are more than 40,000 possible ways to assemble them (remembering the classes of combinatorial analysis: $8! = 8 \times 7 \times 6 \times 5 \times 4 \times 3 \times 2 \times 1 = 40,320$).

In other words, the correct mounting probability is 1/40,320 or approximately 0.0025%.

People less familiar with maintenance routines may be thinking, "What a stupid example! This situation is unreal...."

I first saw this example, few years ago in Health and Safety Executive, n.d.b, and I immediately recalled a situation very similar to the example of the bolt and washers that occurred in a scheduled maintenance intervention that was performed on a tread extrusion line.

The abovementioned line is intended to cool the treads which are produced in a rubber extruder. This line had a roller conveyor called "shrinkage conveyor" and on that occasion one of the various tasks scheduled for the mechanical maintenance team was the replacement of the roller bearings drawn from that conveyor.

Once the maintenance tasks were completed, the cooling line was returned to the production team who started the extrusion line operation.

The first treads produced had a variation in length outside the specified values and this problem caused a new shutdown of the equipment to identify the cause of such variation, since several factors could cause this problem.

After several checks it was found that a novice mechanic had assembled the rollers of the shrinkage conveyor and he had never disassembled such a conveyor, received no special information about the assembly and also did not check any technical documentation to make such assembly and, for this reason, he did not notice that the shrinkage conveyor has this name because the diameters of the rollers are different, but the difference is not recognized by naked eyes.

In short, the shrinkage conveyor is composed of a series of power-driven rollers over which the extruded stock passes. The rollers are successively smaller in diameter, and being driven at the same rpm, their peripheral speed decreases; the linear speed of the stock is thus reduced, and its natural shrinkage hastened before it is cut into tread lengths.

Let's make a reflection on the shrinkage conveyor case. If we apply the root cause analysis to analyze this incident, could we exclusively define the root cause of the problem as human error and close the case?

Certainly not! Several organizational factors influenced the execution of this task, such as supervision failure to designate an inexperienced mechanic to execute the task without proper guidance, lack of training, lack of technical documentation and, especially, improper assembly conditions, causing the error.

Now let's go back to the original example of the "bolt and washers" as the proposed solutions to this problem also apply to the case of the shrinkage conveyor: check the possibility of changing the design to make it impossible to assemble it incorrectly, define a color code, numbers or other marking to determine the correct order of assembly or make sure a second person checks the assembly, if it is critical.

Returning to the Error Management, we will now reflect on how the organization influences human error in maintenance, or in what context the maintenance team operates in the most diverse industrial segments. And the proposal is to reflect on the answer to each of the following questions:

- Is the maintenance function understood and supported at all levels of the organization?
- Is the maintenance strategy perfectly aligned with the company's strategy?
- When critical equipment is subject to maintenance intervention, is sufficient time allowed for the team to perform the repair, including possible tests, or is pressure exerted on the performers to restart the equipment in the shortest possible time, prioritizing time over quality of service?
- Are maintenance and operational needs well balanced?
- Does the company recognize the good "firefighter," that is, the professional that repairs the equipment in the shortest possible time, at any time of day or night, and completely ignores those who work proactively, preventing the equipment from breaking?
- Have all maintenance personnel been trained and qualified for the tasks they perform?
- Is all technical documentation required for maintenance, including software, complete, up-to-date and easily accessible to all who need it, at any time?
- Are the tools and working conditions adequate?
- Are the teams (shifts when they exist) well distributed and the workload well balanced in order to avoid overloading the professionals?

These questions do not have an easy answer, especially if the respondent is someone from top management in the company. Most of the time, the answer will come accompanied by a series of excuses to justify why things are not as they should be.

As we have seen in root cause analysis, without understanding the organizational context, any failure analysis that involves a human failure, as a rule, the group is tempted to stop the analysis on the behavior of the individual; the natural tendency is to concentrate the analysis on who did what, without looking at the underlying problems that propagate their behavior.

I have two striking examples, at least for me, of organizational problems that created an environment conducive to failure, or at least that made it difficult to identify them.

## THE ORGANIZATIONAL PROBLEMS CASES

A novice electrical engineer was hired as head of electrical maintenance in a plant, replacing a professional who had been in the job for several years. On a certain day, as one critical equipment was stopped by an intermittent failure, one of the electricians came to his office asking him to help locate the fault in the equipment's AC/DC drive panel.

He felt that the team, or specifically the electrician, wanted to test his knowledge because he had been hired recently and barely knew the equipment.

After asking the electrician a few questions, he found that the electrician actually needed only the drive's manual to identify the error code on the panel because the person he had replaced in the maintenance management kept all the technical documentation in his office, under keys, claiming that he needed to preserve the physical integrity of the documents.

Whenever some repair of equipment demanded any information, beyond what the person might have in mind, he would have to turn to the boss at any time of the day or night.

Someone may say this is a pathological case.

I think this was cultural because it was not so rare to find insecure people, hiding information as a way to secure their position, trying to be recognized as "the only one able to repair certain equipment" or "the irreplaceable one."

Another case to exemplify how some industrial leaders face maintenance.

Years ago, when the PC Laptops were very expensive, a maintenance manager was responsible for several equipment whose automation was done through programmable logic controllers (PLC) and to transfer, edit and back up its programs, his team used an old portable industrial computer that belonged to the department of projects.

Besides being heavy and slow, the computer was not always available when they needed it and this became a frequent justification for the high repair times on the equipment, when the access to the PLC program was required for troubleshooting.

One day, after an intervention took much longer than it should in a critical equipment due to the laptop's unavailability, the maintenance manager decided to make an investment request to purchase a PC Laptop, exclusive for his team, as the reduction of equipment downtime fully justified the investment.

He asked the IT Department for an estimate cost of the PC Laptop, according to the technical characteristics recommended by the PLC manufacturer. After receiving the estimated value of the computer, he prepared the calculation of the return on investment (ROI), based on the estimated downtime reduction and forwarded the investment request to the industrial manager accompanied by a neat justification.

Contrary to his expectation, his request was approved in record time, without the usual questioning of the industrial manager and other approvers.

One fine day, a computer technician reported that the PC Laptop had arrived and asked the maintenance manager to check the equipment and acknowledge receipt, signing a term of responsibility, which was internal company standard.

When the maintenance manager saw the computer and checked the specifications on the invoice he was surprised because it was not the equipment that had been requested. The one received was far superior to the one originally specified.

He was even more surprised to find that the purchase specification was not the same as his investment request and continued to track until he discovered that the technical specifications contained in the investment request had been changed by the industrial manager, his immediate superior, and corresponded to what was purchased.

Very surprised, the maintenance manager sought the industrial manager to know the reason for changing the specification of the computer and heard him explaining the following:

> "I needed to change my laptop for a better one and, in order not to make another invest-ment request, I 'took advantage' of the one made for maintenance and changed the specification to what I needed, so that I get the new computer and mine, which is a little older and has only intermittent 'minor' problems on the screen that sometimes makes the image disappear, goes to you, that is, maintenance. This way, everyone is supplied with the equipment they need, without bureaucracy and without the need for approval from the top management."

## THE OLD-FASHIONED MAINTENANCE HERO

Maintenance work often drives people out because, intuitively, it has always been associated with unfavorable working conditions such as working under pressure, long working hours, including weekends and holidays, and lack of recognition, among others.

Just as maintenance work drives people away, it also produces heroes, those peo-ple who identify with the conditions mentioned and think that working long hours is "normal" and, as we said before, like to be "firemen," standing out in the group because they are professionals who repair the equipment in the shortest possible time, 24/7 at any time of day or night.

In addition, they are proud to be the only ones repairing certain equipment and what is worse, they think that this way of acting is the best way to contribute to the company where they work.

What causes more concern is that some companies still promote, directly or indi-rectly, this type of attitude because they work exclusively in reactive mode and, in this way, the "heroes" are those who repair the equipment in the shortest possible time, although in most cases, sacrificing the service and even safety.

According to studies conducted in 2000 in the Unites States of America, the maintenance of American surveyed companies worked more than 55% of the time in the reactive mode.

We have mentioned before, but it is worth remembering again the inconveniences of working in the reactive mode or break-fix. These include the following:

- Emergency corrective maintenance increases chaos and confusion, com-promising quality of service execution and safety of performers.
- Increases manufacturing cost due to unplanned shutdown of critical equipment.
- It increases the cost of maintenance work, especially if overtime is required.
- In many cases, it requires purchases and hiring on an emergency basis leading to additional costs, mainly transportation costs, and not always the amount paid for the item purchased will be the lowest.
- Inefficient use of maintenance workforce.

In summary, maintenance professionals should keep in mind that the new "hero" of maintenance is one who proactively works with safety and quality, avoiding unplanned failures in equipment.

I used to tell my team, in the opportunities in which I managed the industrial maintenance, that companies do not need (and do not want) martyrs or heroes. They need people who are normal, healthy, well prepared, dedicated and committed. And for a person to be healthy, he needs to have a balanced life with enough time for leisure and dedication to social and family life.

For me, people who live exclusively for the company, who work and virtually "live in factories" are sick, and these people do more harm than good to the organization, especially if they are managers and want their employees to behave the same way to be evaluated as "committed and motivated." With this preamble we are now going to address the maintenance managers.

## THE MAINTENANCE MANAGERS

The maintenance manager is responsible for all the technical aspects of the function, but he also has an important role with the company's senior management. Increasingly, maintenance managers need to adapt to the new business reality and not only communicate through technical language but must also use language that is fully understood by executives and those who make the strategic decisions of the company. In most cases, to be perfectly understood by interlocutors of this level, the technical language needs to be translated in terms of costs, risks, return on investment, value added, etc.

The maintenance management function must relate to the following levels of company:

- Factory floor: encompasses the management of the execution of the maintenance itself including methods, processes, plans, quality control, safety and analysis of operational performance indicators.
- Maintenance department: economic and strategic management by defining the maintenance strategy, including all aspects of maintenance and reliability engineering, monitoring relevant performance indicators, management of human and material resources, subcontracting policy, legal and regulatory aspects, etc.
- Executive level: participation in the definition of the corporate plan for the management of physical assets with regard to the policy for the replacement of assets at the end of their useful lives, acquisition of new equipment, development of human resources, etc.

If I am asked, based on my personal experience as a maintenance manager and administrator in other areas, one of the most important skills of managers and leader is, without a doubt, communication. In a way, the other characteristics, essential for a manager, to be correctly applied in management, will always depend on excellent communication.

For example, can one imagine a leader motivating or guiding his commanders if he cannot communicate properly with them? Or how a leader can influence people to believe and follow the established strategy if they cannot make themselves understood?

On communication, I remembered a maintenance management assessment that I did some time ago in one of the South American countries, and I found it strange how the maintenance personnel referred to the industrial manager, they called him "dog without a tail." Not containing my curiosity, I asked one of the people what was the reason for this nickname (obviously the industrial manager himself didn't know he had this nickname) and the answer was that the manager would never show his feelings and, in practical terms, they never knew if he had been satisfied or not with some service carried out by the maintenance team, for example. In other words, he didn't "wag his tail" as a pleased dog would.

Still about management styles and management problems, I will transcribe an article that I wrote about one of the problems that usually affects maintenance, the micromanagement, an evil afflicting organization and destroying careers.

What is micromanagement? The definition of micromanage can be "managing with excessive control or attention to details". At this point, I am sure many will have in mind some manager who closely monitors his/her team, wants to know of the smallest details, especially what his/her team is doing, doesn't delegate, doesn't let anybody decide anything and mainly performs personally some tasks that would be the responsibility of the team. This last characteristic gives these managers the title of "hands-on," a title that some even pride themselves on owning.

As I spent much of my career working in industrial maintenance it was in this universe that I found some examples of managers who spent most of their time on the "shop floor" together with the team, literally performing the tasks that would be the responsibility of their subordinates.

Obviously, I have, on many occasions, accompanied maintenance teams that I managed in critical interventions; in maintenance, this is inevitable but, it was said, the success is in the balance.

If someone asks me, after all, what is the rule? The answer is simple: common sense!

There are critical interventions, the failure of which will lead to risks to the safety of people, the environment or great losses, in such cases, even as support for the team, the presence of the manager is important and necessary. In fact, what we call micromanagement is the behavior of the manager who spends most of his time with the team, accompanying any task and, most of the time, ends up bringing loss to the organization and discouraging its personnel.

Usually they are professionals who have been promoted without the proper preparation and here prevails the saying that goes "a good technician will not necessarily be a good manager." Some people feel much more comfortable doing what they were accustomed to perform than to risk doing new tasks, which are really their obligations; preferring to stay in the so-called "comfort zone."

Some years ago, I wrote a short story to illustrate micromanagement, trying to show some managers how this kind of attitude can be very damaging to the company and especially to their own careers. Here it goes.

## THE FRUIT FARM: A TALE ABOUT THE MICROMANAGEMENT CURSE

Once upon a time, a farmer decided to begin growing fruit trees. After some time, he felt that it was necessary to hire somebody to help look after the trees.

For every tree, the farmer would hire one person to watch the tree grow. While it would be important to help pick the fruit, the employee must also watch out for pests, insects, bacteria and fungi, all of which could affect the fruit's health.

Employees would perform the daily tasks and, more importantly, keep a close eye on the growing fruits. The employee would be responsible for the entire tree.

By the time the fruit trees became an orchard and more employees were hired to watch the trees, there was a need for a new position: a supervisor.

The supervisor would be responsible for the whole orchard instead of only one fruit tree.

He was positioned at a higher level to oversee the orchard. Yet, he could not see the same fruit details as the other employees…

The business continued growing, and the orchard became a fruit farm. Once again, there was a need for a new position.

Now the farm needed someone to conduct business affairs. This new employee would provide direction, guide other employees, administer and organize work process and systems, and handle problems. A manager was needed.

This person was positioned in a superior level in order to have a bird's eye view of the farm and allowed him to see outside the farm borders, both opportunities and threats.

The manager was in charge of the strategy of the business and looking towards the future…

Everything was going well until the day the manager decided to go down to help the supervisor. And then the supervisor, in turn, had gone down to help the employees.

But no one was watching the future of the business because the manager and supervisor were too busy looking at single trees while the farm borders were unattended.

One windy day, a fire in the neighborhood crossed the farm borders and destroyed the plantation. Nobody had been watching and, consequently, nobody took actions at the right time. The fire brought the farmer to ruin and all the employees lost their jobs.

The manager forgot (or didn't know) that the essence of management is not centered on accomplishing tasks, but growing results. The manager needed to allow his employees to complete the tasks so he could concentrate on the future vision.

# 7 Other Important Building Blocks of Maintenance Management

## ORGANIZATION OF MAINTENANCE

One question that I have often heard is: "What is the most efficient organizational structure of maintenance?" And the answer has always been the same: The most efficient organizational structure is not theoretically the best but the one that is applied correctly. And the reason for this type of response is that, in practice, we find many distortions in terms of the organizational structure of maintenance for a variety of reasons, which have nothing to do with the real needs of the company, among which we can mention: personal vanity, political influence, power play, etc.

At the beginning of implementation of the maintenance process, it is recommended to adopt a simple structure where it is clear to all the members of the team their roles and responsibilities and the processes with definition of inputs, outputs, information flows and interdependencies.

Before we look at the different types of maintenance organizational structures, we will detail the functions necessary for the maintenance department to properly perform its role. As such, maintenance manager: Responsible for maintenance management as detailed in Chapter 6.

Maintenance and reliability engineering: Provides leadership, direction and technical expertise to optimize and sustain reliability, maintainability and Life Cycle Cost (LCC) of equipment and facilities. While maintenance engineering can't affect performance directly, it provides the data needed to optimize maintenance and develop strategies that support continuous improvement. Among its activities we highlight: Root Cause Analysis (RCA) including failure analysis and reporting, reliability analysis (life data analysis—LDA, reliability and maintainability analysis—RAM, etc.), analysis of inspections and condition monitoring, evaluation of the integrity and extension of the useful lives of assets, prioritization of improvement projects, return on investment—(ROI) calculations and Life Cycle Cost—LCC, criticality analysis and its application, definition and application of maintenance strategy and policies, preparation of maintenance and inspection plans, measuring and improving the productivity of the maintenance team, elaboration and qualification of personnel in technical maintenance procedures, definition of methodology and tooling required to perform maintenance tasks, elaboration of improvement projects in equipment and facilities, identification of spare parts and recommendations for stock levels.

Maintenance Planning and Scheduling: These are the processes governing the execution of maintenance, including the Work Order (WO) management process from planning to the scheduling of WOs executed. This subject will be discussed with more details later in this chapter.

Material and Spare Parts Management: Responsible for the management of spare parts and materials including registration, maintenance of inventory levels, replacement, storage and delivery to users.

Execution: Team of professionals who effectively perform maintenance tasks, including the various required specialties such as mechanics, lubrication, electrical, instrumentation, boiler and welders, acting in all types of maintenance (preventive, corrective, predictive, etc.).

## TYPES OF MAINTENANCE ORGANIZATION

There are basically two types of maintenance structure.

Centralized structure: Organization is based on several specialized maintenance teams and usually serves a whole site. The centralized organization usually has a mechanics department, an electrical department, etc. As benefits of this structure we can mention: better prioritization and use of resources, better use of qualified personnel and a better job preparation, more efficient use of special equipment and devices, better spare parts control, better control of subcontracted work, improvement of quality and efficiency of maintenance work through the centralization and the supervision structure can be much leaner.

As disadvantages we can mention: Those involved in maintenance are scattered throughout the facility, making it extremely difficult to supervise, greater travel time and, possibly, higher transport costs between plants, the time interval between issuing a WO and its execution may be longer and conflicts between production and maintenance can occur due to priorities (Figure 7.1).

Decentralized structure: The organization is structured according to the layout of the plant and the production process areas. Each production area has a group responsible for its maintenance, and this group can also carry out maintenance tasks as well as planning and scheduling and, in some cases, also managing spare parts. Normally, maintenance and reliability engineering are centralized, even in the decentralized structure. The benefits are faster response to unit requirements, improved interaction between operation and maintenance personnel, less displacement for maintenance personnel, maintenance personnel has greater knowledge of the equipment and failure modes of the area and more effective supervision.

The disadvantages are the execution of larger and more complex services becomes more difficult, tendency to have more staff than is effectively needed when there is no synergy between areas, duplicity of special tools and instruments, increased difficulty in developing staff training and qualification programs, the specialists are used inefficiently, and this tend to demotivate these professionals (Figure 7.2).

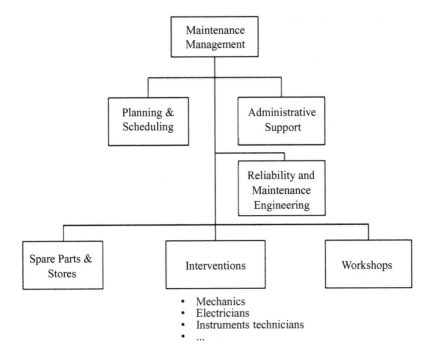

**FIGURE 7.1**  Centralized maintenance structure.

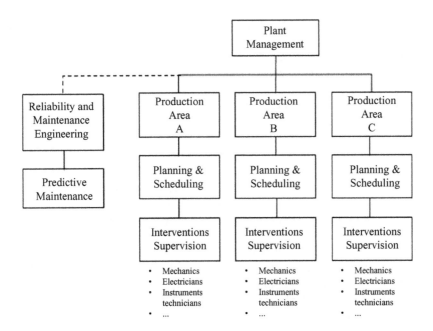

**FIGURE 7.2**  Decentralized maintenance structure example.

## PLANNING AND SCHEDULING

Let's imagine the situation of a maintenance department that we described in Chapter 3, where the overloaded team was only dedicated to the emergency corrective maintenance, that is, working only to "put out fires." Another characteristic of those maintenance departments, that we have not yet mentioned, is the total lack of planning.

The main reason behind the failure to plan is that, in these places, under the described circumstances they work, the exclusive activity of "putting out fires" is considered a priority over any planning.

When we asked the maintenance managers who live in this situation the reason for not having adequate planning, the most frequent justification is that they do not have time to plan because there are so many failures and everything is urgent.

We could argue: If you do not have the time to do it right, would you have the time to do it over and over again?

As we said earlier, reactive maintenance simply becomes a vicious circle and it will require time, efforts and a mainly new mindset to change the situation!

## THE PLANNING PROCESSES

Planning is, of course, the beginning of the maintenance management process. The purpose of maintenance planning is to make the applicable preparations to allow maintenance activities to be performed properly with safety and quality, using the appropriate resources at the right time and cost.

Maintenance planning consists of detailing and documenting the "how to" service in terms of personnel, spare parts, tools, risk assessment, safety precautions, equipment status and necessary documentation (diagrams, drawings, manuals, etc.). In summary, all the information needed to perform a well-defined sequence of operations.

We could say that a maintenance WO is well planned when it, basically, has the following characteristics:

- The right professionals: Properly qualified people with the knowledge and experience necessary for the scope of work. In addition, the WO form should contain the number of professionals required to perform the service.
- The exact location: Precise identification of the process, equipment and location where the service will be performed to reduce travel times, research and adjustments.
- The right time: How long will it take to complete the job. This will help the programming process, which should reconcile with the production expectation regarding the beginning and conclusion of the work.
- The correct parts, tools and equipment: All spare parts must be identified, ordered and available prior to service scheduling. This will help the scheduling process, which should reconcile with production expectation when starting and completing the job. Preparations should be made, so that special tools and ancillary equipment are also available for use when needed.

- Technical specifications, work permits, risk assessment, work instructions and appropriate documentation. All necessary documentation should be available as an integral part of any work package.

These elements are critical for the work to be completed within the planned time with minimum delays, quality and safety.

## THE SCHEDULING PROCESSES

The maintenance scheduling, in brief, consists in determining when the service will be executed. While the objective of planning is to reduce delays, once work has begun, the scheduling aims to eliminate delays among the various services, ensuring that there is enough work with the correct priority, allocated to the available workforce, ensuring this way, the best use of available hours. In this way, the activities of the scheduling consist of the following:

- Obtain the total available working hours by days for the crew;
- Remove administrative hours for each day: vacations, holiday hours, training hours, meeting hours, etc.
- Schedule preventive maintenance task hours for each day;
- Use the remaining hours left for each day to schedule backlog WOs (planned corrective maintenance);
- Generate, in this way, service scheduling (usually weekly);
- Organize the weekly meeting of work scheduling with representatives of maintenance and operation to consolidate the work schedule for the following week;
- Register in the computerized maintenance management system (CMMS) the data from the WOs, which will allow to assess the level of planning and completion of the schedule.

Work scheduling should consider: Service priority and this priority classification should be formally established between maintenance and its customers, especially Operations to avoid everything being "emergency" where a criterion for maintenance planners can't be established, the availability of resources for execution and the release of equipment or facilities by Operation (Table 7.1).

It is up to the maintenance planner to perform the calculations that match the workload with the availability of resources, issuing a preschedule of the jobs to be executed the following week.

In the weekly scheduling meeting, the Planner and the maintenance supervisor consolidate with the Operations the list of services that will be effectively performed.

## BACKLOG OR FUTURE WORKLOAD

The backlog can be defined as a measure of the remaining work to be accomplished, that is, at that instant, if no new WO is received and approved, how long it takes to complete all pending services with the current team.

**TABLE 7.1**
**Typical Example of Work Order Prioritization**

| Priority | Description | Response |
|---|---|---|
| 1 | Emergency<br>• Involves imminent risk of accidents or damage to people/installations.<br>• It causes total or partial production stop, with major implications in the production plan. | Highest priority level. Work must be performed immediately. |
| 2 | Urgency<br>• Involves production-stop risks and implications in the production plan.<br>• It causes great reduction of operational reliability, or of the security system of installations, or variations in the quality of the product. | Work to be completed during the week that cannot wait until the next week to be scheduled. |
| 3 | Routine with scheduled date<br>Work that have a deadline date to be completed. | Execution with scheduled date. |
| 4 | Routine<br>Work without impending influence on the production plan. | Work can be performed on a planned basis, meaning that it will be performed the next week or later depending on material, labor, equipment/tools availability. |

Usually the backlog is calculated in men-hours to run them or weeks. The backlog can also be considered as a measure of the current manufacturing risk. It is worth mentioning, however, that a minimum level of backlog is interesting because it ensures that there is a good occupation rate of the workforce.

## BACKLOG CALCULATION

The backlog can be calculated (in weeks) by the following formula:

$$\text{Backlog} = \frac{\text{Unscheduled open WO (hours)}}{\text{Work capacity (hours per week)}}$$

Let's see an example of the backlog calculation for a mechanical maintenance team with the following characteristics: Mechanical maintenance team: 10 professionals—Vacations per employee: 120 h/year—Holidays per employee: 80 h/year—Total unscheduled WOs: 1,000 h—Annual preventive maintenance workload: 6,000 h—Average of emergencies/urgencies: 15%

$$\text{Backlog} = 1,000/(10*40*(1-0.15))-(10*120/52)-(10*80/52)-(6,000/52)$$
$$= 5.37 \text{ weeks}$$

Figure 7.3 is usually used to explain the meaning of the backlog and its form of control, comparing it with a water supply system.

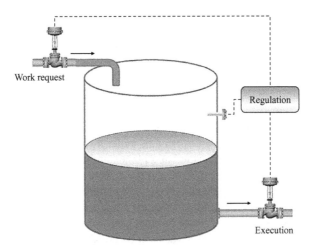

**FIGURE 7.3**    Analogy of backlog control with a water supply system.

The water stored in the water tank represents the volume of services to be performed by the maintenance team of a particular specialty. If at any given time, there is no water inlet and outlet, the remaining volume in the water tank will be the backlog at that particular moment.

The entrance of water represents the new services requested and, consequently, exit represents the completed services.

Observe the control valves installed in the inlet and outlet pipes; they represent the available means the maintenance management has to control the backlog. He or she can reduce the water volume in the tank either by increasing the water flow in the tank outlet or by reducing the water flow in the tank inlet.

It is important to note that before taking any action to reduce the backlog, the maintenance manager together with his or her team needs to do a backlog review; what services are outstanding and why they have accumulated. This subject will be discussed in more detail below.

As for the available means, in the figure represented by the control valves regulation, we can mention:

- Reliability: The best way to reduce the backlog is to eliminate the need to perform corrective maintenance. This can be achieved by increasing the equipment reliability and, as consequence, reducing the number of failures.
- Labor: It is important to check maintenance productivity, to verify that the qualification of labor is adequate for the type of service to be performed. In cases where the backlog is high, for any unspecified reason that requires a specific action, one may consider hiring temporary labor or third parties to perform the pending services.
- Work methodology: Check the quality of planning and preparation of services, especially regarding spare parts and materials, auxiliary equipment (scaffolding, hoisting and loading, etc.) as well as the distribution and supervision of services.

## BACKLOG REVIEW

By principle, the planners should know the current status of each job in the back-log but, in any circumstances, periodically (and especially when the backlog of a particular specialty reaches high levels), it is necessary to evaluate the open WOs because, depending on how the CMMS is utilized, the backlog can be reduced by just eliminating some WOs that should not be considered in the backlog calculation. For example, open WO in CMMS but the job has already been completed but nobody has closed them out, duplicate WO with jobs under different names, WO is pending but the service can only be performed at a plant shutdown and should have another type of treatment, the service, for whatever reason, is no longer required and WO can be closed, service execution depends on parts or materials not available.

After the backlog review and consequent "cleaning" of the open WO in the CMMS, only then the backlog level by specialty can be analyzed. Typically, the backlog should be between 2 and 3 weeks to be considered ideal. If the value is in the range of 4–6 weeks, it is considered critical and over 6 weeks, it will be considered out of control and this is a strong indication that the maintenance department is working predominantly on the reactive mode and in this case, actions should be taken by management to increase work capacity or to reduce service demand as previously discussed.

## WO PROCESS FLOW

WOs are the main document of a planning and scheduling process, because they define and detail the services to be executed, indicate the necessary means and resources, receive the appropriations that feed the financial system of the organization and provide the necessary data to the maintenance history, among others.

From initiation to closure, each WO should be handled according to a standardized process—the WO flow.

The WO flow should describe precisely the logical succession of tasks, who is doing each of them, what is the input and output of each task, which conditions should be set up before executing each of them, and how the WO status evolves along the process.

Because the maintenance WO flow can be recognized as a central process for the maintenance activity involving different functions, the best way to define or review it and get greater commitment from all stakeholders is to apply the process-mapping model.

In the past, when I participated in the CMMS implementation in some places, one of the biggest difficulties we had was to change the local culture of informality and convince the different stakeholders, mainly Operations, to complete the maintenance work request forms.

Nowadays, this problem, if it has not been totally eliminated, we can affirm that it should be minimal. One of the means that helps maintenance in the implementation of the process is the establishment of an official flow chart of the Maintenance Service Order that must be followed by all involved from the beginning to the end of it (Figure 7.4).

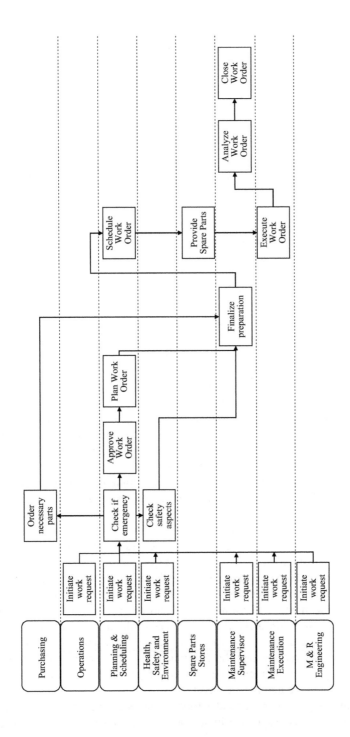

**FIGURE 7.4** Example of WO flow.

A crucial point in implementing the proper planning and scheduling process is the perfect match between three functions represented by the maintenance planner, maintenance supervisor and the operations supervisor.

I will present a real case to illustrate this situation.

## THE PLANNING AND SCHEDULING CASE

In a given plant, after outsourcing all maintenance, it was decided jointly by the maintenance company and the contractor manager that the planning and scheduling process implementation should be reviewed and that the WOs should be scheduled weekly and on a given day of the week there would be a scheduling meeting with the planners, maintenance and operations supervisors to determine which services should be executed the following week.

In theory, only emergency services, according to criteria similar to those presented earlier in this chapter, could cause the services schedules to be modified so that they were executed instead of those that had been scheduled for execution.

There were briefings and awareness-raising meetings as well as training on new practices for all involved, and apparently everyone agreed that the planning and scheduling process, properly implemented, would bring benefits to all involved parties.

In practice, however, this was not exactly the case because the operation's supervisors for many years worked without formalization and were accustomed to simply approaching a mechanic in the maintenance shop and requesting the execution of any service and only afterwards open a service request.

Everything was done in an atmosphere of friendship and camaraderie since everyone had known each other for a long time and apparently worked very well except that it was extremely unproductive, inefficient and much more expensive than it should be.

The operations supervisors felt diminished by having to comply with the bureaucracy imposed by the maintenance contractor and often wait for more than a week for a small service to be performed. It didn't make sense to them! And convinced they were right, they used the excuse of customers to boycott the service schedule as they pleased, to the frustration of maintenance planners who saw their work being totally wasted, and this was reflected in the very low rate of the schedule compliance.

Faced with this situation, the contract manager met with the plant manager and clearly stated the problem and stated that the objectives they had jointly set, involving maintenance performance and the respective performance of the plant, would not be achieved without the planning and scheduling process working properly. The contract manager also reinforced the importance of the mentioned process: Studies show that 3–5 h of execution time can be saved for every hour of advanced preparation, planned jobs require only half as much time to execute as that needed for unplanned jobs and the maintenance plans have direct influence over the production capacity output and the maintenance costs.

In this way, the plant manager understood and agreed with the maintenance man's arguments and assumed responsibility for the success of the planning and scheduling implementation, emphasizing to his employees, mainly to the operations supervisors, that the schedule compliance indicator would be attended by him personally and that he would check all WOs classified as "emergencies" that would affect the scheduling compliance.

To avoid this kind of issue, it is interesting to establish in advance what are the responsibilities of each of the team members (maintenance and operations) to achieve the expected results from the planning and scheduling process. For example:

Maintenance planners: Generally, we consider them the professionals responsible for planning and scheduling WOs. The main function of an individual or group of individuals is to optimize the use of the maintenance workforce through the best practices of planning and scheduling. Their typical responsibilities include to work closely with all supervisors and other maintenance team members, in addition to the supervisor or representative of the operation, plan, document, print and close the WO, create, print and distribute a weekly schedule based on the availability of staff resources, manage the kits and parts that will be used in the WO scheduled to be executed, contribute to the improvement and optimization of the maintenance program, manage maintenance documentation and ensure accurate data collection for analysis, preparation of performance indicators and preparation of reports.

Maintenance supervisors are responsible for conducting their team of professionals in a safe and efficient manner through the execution of the planned and unplanned maintenance jobs. The typical responsibilities of a maintenance supervisor include to work closely with planners, supervisor or operator representative, and other maintenance team members, assign and manage planned, scheduled and unplanned WOs for execution, ensure good communication with operations, review and track daily the weekly maintenance schedule and the quality of the work execution, collect, review and approve completed WO, question whether WO classified as "emergency" or "urgent" can't be performed as planned and scheduled, participate in planning and scheduling activities as required.

Operations supervisors (or any other function assigned to this activity): Their responsibilities with respect to planning and scheduling are communicate daily with maintenance supervisors, coordinate production needs with regard to maintenance, define work priorities for planning and scheduling based on critical equipment and production priorities, act as the link between operations and planning and scheduling, carefully evaluate the weekly maintenance schedule before approving it for execution, ensure availability of equipment for the execution of scheduled maintenance.

## Performance Indicators of Planning and Scheduling Activities

It is very important that the planning and scheduling activities are monitored through performance indicators so that actions are taken to improve or correct some abnormality.

Among the performance indicators typically used in companies in general, we highlight the following and their values are considered Best Practices based on Wireman, 2004 (Table 7.2).

The CMMS is an important tool to support the planning and scheduling process and, when properly implemented, has several features and has numerous advantages over the management of maintenance done in a manual way.

**TABLE 7.2**

**Performance Indicators of Planning and Scheduling Activities**

| Indicator | Value (%) |
|---|---|
| Reactive hours/Total hours | <10 |
| Maintenance schedule compliance | 95 |
| Planned maintenance work | >80 |
| PM/PdM hours/Total hours | 50 |
| Preventive maintenance compliance | 100 |
| Overtime | <5 |
| Work order coverage | 100 |
| Productivity rates (Wrench time) | 60 |

It is very important to emphasize that CMMS is an important tool, but nothing more than that, and it is wrong to believe that only investing in the acquisition and implementation of this system, without qualified professionals and well-defined processes, can improve maintenance management.

This frustration is often revealed when some companies seek out young professionals with a good computer knowledge to work in planning and scheduling without requiring knowledge in maintenance processes, as if software knowledge would be enough for someone to do good planning or scheduling. In other words, the focus should not be software but maintenance processes.

One factor that usually contributes to the maintenance departments of countless companies not being able to enjoy the benefits of CMMS is the poor implementation of the system. We can mention some of the most commonly encountered implementation problems: The implementation plan is superficial without clear definition of the goals and objectives that are to be achieved, deadlines out of reality and sub-sized resources, lack of sufficient resources to implement specific tasks, the implementation project leader (usually from the IT area) doesn't have the authority and backing of the senior management of the company or even without full knowledge of the maintenance processes that will be covered by the CMMS.

Before finalizing this subject, I would like to address a topic that has already yielded some polemics, in my view, for no reason whatsoever which is calculating the productivity of maintenance workers by calculating WO/employee.

## THE MAINTENANCE PRODUCTIVITY MEASUREMENT CASES

**Case #1:** A company wanted to optimize its maintenance management and they hired an internationally renowned consultancy to assess their maintenance processes to identify the opportunities for improvements.

After the assessment, the consultancy concluded, among other things, that the number of maintenance workers could be reduced and, for reasons that are not worth discussing here, I was involved in this case and they asked for my opinion about the consultancy recommendations regarding the maintenance reduction.

I asked them to present me the applied methodology which led them to conclude that there were an excessive number of workers, and to that end, one of the team's consultants was appointed to talk to me.

At the meeting, a very young and arrogant gentleman appeared, speaking very poor English and shortly after the formalities of the presentation, he abruptly questioned me as if I was interested in discovering some secrets or magic formula of his company.

I quickly explained that I had experienced several cases of maintenance productivity measurement using the work sampling methodology and just wanted to know what the theoretical background was that led them to the conclusion they presented to the customer and highlighted that I was not interested in knowing the details of the calculation, much less in discovering any secrets of his company or of any other.

Demonstrating much ill will, the young man said that after some surveys that they did during a certain period, they had done calculations involving the number of WOs per employee and concluded that the nonproductivity was quite significant and that it could be reduced through improvement in planning, enabling the reduction of some workers.

I asked him if this conclusion was for some specific specialty and had been calculated for each of the specialties existing in the plant (mechanical, electrical, instrumentation, etc.) individually.

The gentleman remained silent and, judging that he had not understood, I repeated the question.

Demonstrating a lot of annoyance, as someone forced to train an animal, he replied that they added the WOs of all the specialties together and found the average.

At this point, before he continued the explanation, I gave up and allowed myself to leave the meeting, leaving the young guy speaking to himself...

**Case #2:** Another company started measuring the productivity of their maintenance crew by calculating the number of WOs performed per employee and worse, they spread the results of the newly created indicator to the whole organization, exposing the unfortunate ones who made fewer WOs than others in a given period, classifying them as unproductive.

And, incredible as it may seem, there are a lot of people around the world doing productivity calculations in a similar way.

The last time I had the opportunity to discuss this matter, not long ago in Europe, when I participated in a team responsible for, among other things, setting up an indicator to measure maintenance productivity, I have argued the following:

It is impractical to add WOs from different service scopes because they will have different execution times and the distortion will increase even more if it is done as the consulting firm of Case # 1 has done, adding up hours of different specialties. Let me explain by the following example:

Electrician John performed three WOs on week 14 being:

WO #1—Electrical specialty—Scope: change 20 lamps in the main office. Execution time: 5 h

WO #2—Electrical specialty—Scope: repair in 20 spare lighting fixtures. Execution time: 8 h

WO #3—Electrical specialty—Scope: scheduled repair in electrical conduit pipes in the production line X. Execution time: 18 h

The electrician Peter executed only one WO in week 14: WO #4: electrical specialty—scope: overhaul on the Z-Line electrical panel consisting of the substitution of wiring, contactors, protections, lamps and buttons. Execution time: 34 h

In the reasoning of the one who defends the WO/employee indicator, in the example above, John is three times more productive than Peter, but it is not so difficult to understand that different tasks can't be compared. In addition, maintenance is not serial production, or it is seldom if you have repetitive tasks that make it possible to establish an indicator with these bases.

Accordingly, in the example above, if hypothetically, the week has 40 work hours: John reported 31 h in the week, while Peter reported 34 h. In this case, Peter reported more hours but still can't compare productivity in different tasks.

Back to the discussion I had in Europe, I insisted that one of the established ways of measuring maintenance productivity is the work sampling and someone who strongly advocated the use of the WO/employee index argued that this methodology is very laborious and that there were situations where the WO/employee index could be applied with some caution and to close this matter once and for all I mentioned the famous statement: "there is always a well-known solution to every human problem—neat, plausible, and wrong" (Mencken, 1920).

## SPARE PARTS AND MAINTENANCE MATERIALS MANAGEMENT

Management of spare parts is of utmost importance for the management of maintenance because when poorly executed it can adversely affect the availability of the plant since the lack of spare parts, when needed, causes a considerable increase in equipment repair times. On the other hand, the value of poorly planned inventory can reach significant values and they can negatively affect the maintenance cost.

Typically, companies are experiencing an internal conflict where, intuitively, the maintenance department would like to have the largest possible stock of spare parts to reduce the risk of long downtime of critical equipment due to unavailability of some item and, the financial department faces the problem to reduce capital investment held captive represented by spare parts in stock. This type of conflicting interests and the losses that wrong decisions can bring to the company, reinforces the vital importance of professional spare parts management in any organization.

Among the problems that organizations, in general, face to manage spare parts, we highlight the following:

- Difficulty in determining which parts should be kept in stock as well as the required quantity because a component failure can occur both due to wear by use and randomly by any other cause, as we described in detail earlier, when discussing the component failure probability curves.
- Unavailability of some necessary parts, mainly imported parts or specific parts of equipment, many of these parts being exclusively supplied by the equipment manufacturers or manufactured to order with long lead times.

Based on the above, we can affirm that the goal of spare parts management is to ensure the availability of spare parts for equipment and facilities maintenance, when necessary, at an optimum cost. In addition, such management must ensure that the quality of the parts is adequate, from the acquisition to the use of the parts and this implies maintaining efficient processes of acquisition and storage.

The choice or optimization of a suitable inventory control model is affected by: the nature of the spare part (category), the level of consumption (high or low consumption), the regularity of the demand for the spare part, the relationship between the consumption of the part and the form of maintenance for which it is intended and the end use of certain types of spare parts.

Among the many necessary actions to ensure an effective management of spare parts, we highlight definition of stock parameters, standardization of spare parts description, physical organization and storage requirements recommendations, and the warehouse management of spare parts and materials.

Due to the mentioned circumstances, characteristics of the maintenance, besides the implantation of the system, it is necessary to evaluate the standardization of the inventory items in order to reduce the number of items in stock.

It is also worthwhile to evaluate the possibility of having a contract with suppliers for the high consumption items or even consigned inventory in order to have a low risk of unavailability of spare parts with less capital invested in stock.

## STANDARDIZATION OF SPARE PARTS DESCRIPTION

Let's imagine the following situation: An equipment stops because a certain component of this equipment has failed and needs to be replaced. Let us also assume that this component is not a common item, on the contrary it is something specific to this type of equipment and the maintenance professional in charge of making the repair needs to identify it to know if there is any part in the spare parts warehouse or, in the worst case, buy a new part.

Readers who have already had the misfortune of working in maintenance and not having an organized spare parts warehouse know how difficult is to desperately need a component to repair one critical equipment and have to begin by describing the component to be purchased by the purchasing department.

In the past, it was very common for maintenance supervisors, either out of ignorance, laziness, haste or any other reason, to deliver the defective part to the buyer along with the purchase requisition where the famous phrase was written: "part acquisition according to the attached sample" and in this way, transferring to the poor buyer the responsibility of identifying the component in order to be able to buy it.

I think it is totally unnecessary to describe the drawbacks of this kind of "process"....

It seems obvious that all necessary spare parts should be clearly described with their essential characteristics so that there is no doubt about their identification. To facilitate this process, organized and well-managed warehouses begin by establishing a coding system for the spare parts they keep in stock.

By classifying and coding all parts, the possibility of duplicating spare parts in stock is minimized and, consequently, the inventory will be reduced. The coding also helps the accounting and the implementation of computerized system (CMMS) for inventory management, and facilitates communication between the involved parties, ending (or at least minimizing) the buyers' drama.

For categories, spare parts can generally be classified either as parts specific to a particular equipment or installation and the parts are specially manufactured according to original equipment manufacturer (OEM) design specifications and are not normally interchangeable with parts of another brand or manufacturer or as commercial parts that meet internationally recognized standards and are interchangeable with other brands of the same part.

It should be noted that it is necessary to establish a method for technical description and coding of spare parts to generate a standardized and unique descriptive identification for each spare part or material. Descriptive standardization is necessary due to the large volume of components and materials on the market with different application characteristics. This optimizes the transactions made with materials, from acquisition, planning, purchase, receipt and storage, to the final use of the materials.

In addition to the spare parts coding, it is also required to encode their physical location in the warehouse, in order to reduce the time spent by the warehouse personnel when they need to find a particular part.

## Physical Organization and Storage Requirements Recommendations

The available space is one of the most important points to consider when implementing a spare parts and maintenance materials warehouse because this factor strongly influences the purchase, storage and service strategy.

Anyone who has tried to get an available area to implement or expand the maintenance warehouse in a plant knows how difficult it is to convince those who decide to free up a useful area and invest in the purchase of spare parts and materials, leaving the company's capital frozen. Even with all the valid arguments, it is not easy at all. Therefore, the space of a warehouse must be very well planned and deployed in order to get the most out of its total area.

A crucial point that directly influences the reliability of the equipment and installations is the correct storage of parts and materials in the warehouse. As we saw earlier, incorrect storage can lead to premature failure of parts and components, and unfortunately it is very common to see materials being stored without the conditions recommended by their manufacturers being followed.

Some time ago, I participated in the implementation of the maintenance system in a company located in a desert region in the Middle East. When I was there, the industrial unit was still in the transition between the termination of installation and beginning of commissioning and many facilities were under construction among which the maintenance spare parts and materials warehouse.

The leader of the maintenance implementation project was worried about the spare parts and maintenance materials that had been purchased together with the equipment and which were being temporarily stored in a shed that, although covered, still had no lateral closure and, in this way, the parts were subjected to very high temperatures and the dust of the desert brought by the wind.

He asked me to visit with him the place where the pieces were stored and the material so that I gave my opinion about the situation.

In a huge shed, with temperature above 40°C and dust all over the place, I noticed many pieces stored in a precarious way and I told my colleague that, in my opinion, the environmental conditions were far from ideal and certainly would be detrimental to those parts stored therein, especially the bearings and precision hydraulic components.

He led me to a large room built inside the shed, which served as an office for the people who were responsible for the materials. When entering the shed the thermal shock was brutal because there was an air conditioning running at full power, leaving the room as cold as a refrigerator, contrasting with the oven from where I had just left.

In this place, I was introduced to the person in charge of the sector who showed me the rest of the materials they stored. Inside this area where we were, with temperature controlled by air conditioning and without dust, they stored personnel uniforms, personal protective equipment (gloves, glasses, helmets, boots, etc.) as well as many bottles of mineral water.

Undoubtedly, from the standpoint of equipment reliability and maintenance cost, there was a clear reversal of values as components and materials, usually of high value, that require special storage conditions, perished under the heat and dust of the desert, while the other materials, which were also important but could for the most part be stored outside the room, were protected from adverse climatic conditions.

When we spoke with the warehouse manager, we alerted him to the fact that many pieces that were stored in the shed could be damaged due to local conditions but, he reacted negatively to our comments and said that the storage was temporary and the definitive storage was still under construction.

The problem is that many pieces exposed to heat and dust, until completion of the installation and final installation of the warehouse, could be damaged when they were finally stored in the appropriate place.

I suggested to the project management leader of maintenance management to make a formal recommendation to the industrial manager warning about the problem and attaching some recommendations for storage of the parts manufacturers who were exposed to heat and dust.

# 8 Epilogue

We reached the end.

I hope I have met the expectations of the readers and I want to use these last pages to, among other things, share with those who do not know the formula of customer satisfaction. Something so important that someone decided to express it by a mathematical formula, I think, to facilitate its understanding by people like us engineers, who seek to identify logical sense in everything. Here is the formula:

$$\text{Custome satisfaction} = \text{Delivery} \times \frac{\text{Perception}}{\text{Expectation}}$$

Analyzing the formula, we understand that customer satisfaction is directly proportional to the product of the "delivery" (what the customer actually receives) by the "perception" (how the customer perceives what is received) divided by the "expectation" (what the client hoped to receive).

Do you believe that the same product (or service) can cause satisfaction or dissatisfaction with the same customer?

The answer is yes and I will justify it; suppose you invite your beloved one to spend a weekend in a 5-star hotel. You describe a luxurious beachfront hotel with beach service, cocktail reception, romantic candlelit dinner on Saturday night and everything anyone can imagine in terms of luxury and comfort.

You take her/him to the hotel and, upon arriving at the front desk, has an unpleasant surprise; due to a problem in the system, your reservation was canceled and the hotel is full.

The hotel manager apologizes for the mistake in service and makes up for it by offering you free stay for the couple in a 3-star hotel of the same hotel chain. The hotel is about 100 meters from the beach, has a comfortable room with air conditioning, cable TV, and other standard amenities, but no luxury.

How would your girlfriend/boyfriend feel? I think most normal people would be very frustrated and dissatisfied in a situation like this.

Now let's imagine another situation; you invite the same person from the previous case to spend the weekend at a campsite near a beach; you say that you are out of money and that the most important thing is not the accommodation but the fact that you are together.

You warn your girlfriend/boyfriend not to forget the insect repellent and the fan because this beach is famous for mosquitoes that attack ruthlessly day and night, without a break.

On the way to the campsite, you change the itinerary and surprise your companion by taking her/him to spend the weekend in the same 3-star hotel of the previous example.

What will the reaction be now? You both will probably be satisfied unless she/he likes to sleep on the floor, bathe in cold water and serve as food for mosquitoes...

So, what has changed? Just the expectation...

I apologize to those who have been frustrated with the reading. I have tried to put in these pages a synthesis of what I have seen, lived, learned and taught in over 39 years of career, traveling in more than 35 countries on all continents, where I managed, supported the implementation of maintenance management, assessed maintenance, provided training for all of these topics covered in this book, thinking that this can be useful for those who are starting their careers in maintenance or for those already in maintenance and want to improve maintenance management and obtain better results.

As a final reflection, I wanted to say that I end my career satisfied by having followed my father's advice regarding my professional life, which was something like this:

"It does not matter what profession you choose; any profession is good as long as you enjoy it and do well everything that needs to be done.

Be honest, straightforward and dedicated.

Keep your focus exclusively on work; do not aim for the money. The reward will come, in one way or another, to the extent of your competence, effort, and dedication.

Do not look for shortcuts because they will never take you where you want to go.

Get away from the political career because the greatest virtues of a politician are lies and falsehood."

I've had joys, achievements and, frankly, I can say that I got where I wanted to go.

In the same way, I stumbled, made wrong decisions and had some frustrations, but this ends up being forgotten and I do not carry regrets or resentments, remembering again the words of my father: "time heals the wounds."

# Bibliography

Bannister, Ken. *From Our Perspective: The Study*. Efficient Plant Formerly Maintenance Technology. www.efficientplantmag.com/2014/12/from-our-perspective-the-study. Willowbrook, December 14, 2014.

Blann, Dale R. *Maximizing the P-F Interval through Condition-Based Maintenance*. MaintWorld Magazine for Maintenance & Asset Management Professionals. www.maintworld.com/applications/maximizing-the-p-f-interval-through-condition-based-maintenance. October 07, 2013.

Doyle, Arthur Conan. *A Study in Scarlet*. London: Ward Lock & Co., 1887.

Doyle, Arthur Conan. *The Sign of Four*. London: Spencer Blackett, 1890.

Doyle, Arthur Conan. A Scandal in Bohemia. The Strand Magazine, 1891.

Doyle, Arthur Conan. The Adventure of the Copper Beeches. The Strand Magazine, 1892.

Doyle, Arthur Conan. *The Hound of the Baskervilles*. London: George Newnes, 1902.

Dunn, Richard L. *Benchmarking Maintenance: When You Start Feeling Good about Yourself, It's Time to Benchmark. That's a Rule of Thumb from Bob Schmalbach, Chairman of the Foundation for Industrial Maintenance Excellence and a Retired Plant Engineering Consultant for Dupont*. Plant Engineering Magazine. www.plant-engineering.com/single-article/benchmarking-maintenance/ec584fbd54e5fae00cb5c-fa1bdd1dfc6.html. Downers Grove, March 01, 2001.

Dunn, Sandy. *Managing Human Error in Maintenance*. www.plant-maintenance.com/articles/Human_Error_in_Maintenance.pdf. Accessed on August 15, 2016.

Eckert, Catherine. *Apollo South America: Apollo Análise de Causa de Raiz (RCA)—Um Sumário*. Santos, 2005.

Grupo WEG. *Motores Elétricos: Guia de Especificação. Código 50032749*. https://static.weg.net/medias/downloadcenter/h32/hc5/WEG-motores-eletricos-guia-de-especificacao-50032749-brochure-portuguese-web.pdf. Jaraguá do Sul, 2017.

Health and Safety Executive. *Human Factors Briefing Note No. 6- Maintenance Error*. http://www.hse.gov.uk/humanfactors/topics/complete.pdf. Under Open Government License. www.nationalarchives.gov.uk/doc/open-government-licence/version/3. n.d.a.

Health and Safety Executive. *Human Failure Types*. www.hse.gov.uk/humanfactors/topics/types.pdf. Under Open Government License. www.nationalarchives.gov.uk/doc/open-government-licence/version/3. n.d.b.

Health and Safety Executive. *The Flixborough Disaster: Report of the Court of Inquiry, HMSO, ISBN 0113610750*. www.hse.gov.uk/comah/sragtech/caseflixboroug74.htm. Under Open Government License. www.nationalarchives.gov.uk/doc/open-government-licence/version/3. London, 1975.

Health and Safety Executive. *Reducing Error and Influencing Behaviour*. https://books.hse.gov.uk. Under Open Government License. http://www.nationalarchives.gov.uk/doc/open-government-licence/version/3/. London, 1999.

Idcon Inc. *Maintenance Dictionary. Fixed Time Maintenance (FTM) Entry*. www.idcon.com/resource-library/maintenance-dictionary.html. 2019.

Ivara Work Smart. *Ivara EXP Supports INPO AP-913*. 2007.

Lafraia, João R.B. *Manual de confiabilidade, mantenabilidade e disponibilidade*. Rio de Janeiro: Qualitymark, 2001.

Latino, Robert J.; Latino, Kenneth C. *Root Cause Analysis: Improving Performance for Bottom-line Results*. 3 ed. Boca Raton, FL: CRC Press, 2006.

Latino, Robert J. *The top 10 Reasons People Don't Trust Root Cause Analysis.* Plant Services. www.plantservices.com/articles/2006/240.html. September 15, 2006.

Leroux, Marc. *OEE as a Financial KPI—How to Explain OEE in Financial Terms to "C" Level Executives and Use DuPont Model to Visualize the Impact that Manufacturing Improvements Can Have on Financial Performance.* ABB. https://new.abb.com/cpm/production-optimization/oee-overall-equipment-effectiveness/oee-as-a-financial-kpi. n.d.

Mencken, Henry Louis. *Prejudices: Second Series.* p. 155. New York: Alfred A. Knopf, 1920.

Mobley, Keith. *Maintenance Engineering Handbook.* 7 ed. New York: McGraw-Hill Education, 2008.

Moubray, John. *Reliability-Centered Maintenance.* 2 ed. Oxford: Butterworth-Heinemann, 1997.

NASA. *NASA Reliability and Maintainability (R&M) Standard for Spaceflight and Support Systems STD-8729.1A.* Houston, 1997.

Nascif, Júlio; Dorigo, Luiz Carlos. *Manutenção orientada para resultados.* Rio de Janeiro: Qualitymark, 2009.

Palady, Paul. *FMEA: Análise dos Modos de Falha e Efeitos: prevendo e prevenindo problemas antes que ocorram.* São Paulo: IMAM, 1997.

Pallerosi, Carlos. *Confiabilidade: a quarta dimensão da qualidade.* São Paulo: ReliaSoft, 2007. V. 1 and 10.

Palmer, Richard. *Maintenance Planning and Scheduling.* 2 ed. New York: McGraw-Hill Education, 2006.

Porril, David. *Constructing an Effective Maintenance Plan.* Reliable Plant. www.reliableplant.com/Read/338/effective-maintenance-plan. n.d.

Reason, James; Hobbs, Alan. *Managing Maintenance Error. A Practical Guide.* Hampshire: Ashgate, 2003.

Reliasoft. *Reliability Basics—Relationship between Availability and Reliability.* Reliability HotWire—The e-Magazine for the Reliability Professional. www.weibull.com/hotwire/issue26/relbasics26.htm. April 26, 2003.

Suzuki, Tokutaro. *TPM in Process Industries.* Portland, OR: Productivity Press, 1994.

United States Department of Energy. *Root Cause Analysis Guidance Document. DOE-NE-STD-1004–92.* Washington, 1992.

United States Military. *Military Hadbook: Electronic Reliability design Handbook 338B.* Washington, DC: D.O.D., 1998.

United States Military. *Reliability-Centered Maintenance (RCM) Process. Mil. Std. 3034.* Washington, DC: D.O.D., 2011.

United States Military. *Failure Modes, Effects and Criticality Analysis (FMECA) for Command, Control, Communications, Computer, Intelligence, Surveillance and Reconnaissance (C4ISR) Facilities.* http://www.wbdg.org/ffc/army-coe/technical-manuals-tm/tm-5-698-4. Accessed on August 22, 2016. Washington, DC: D.O.D, 2006.

Wireman, Terry. *Maintenance Basics: The First Step to Achieve Zero Breakdowns.* Ridgefield, CT: Genesis Solutions, n.d.

Wireman, Terry. *Benchmarking Best Practices in Maintenance Management.* New York: Industrial Press Inc., 2004.

Villaran, M.; Subudhi, M. *Aging Assessment of Large Electric Motors.* Nuclear Power Plants—NUREG/CR-6336. Washington, 1996.

# Index

Milton Keynes UK
Ingram Content Group UK Ltd.
UKHW040051071024
449327UK00019B/470